Junior Skill Builders

GEOMETRY in 15 Minutes a Day

LEARNINGEXPRESS®

NEW YORK

Library of Congress Cataloging-in-Publication Data:

Geometry in 15 minutes a day.—1st ed.
    p. cm.
  Includes bibliographical references and index.
  1. Geometry—Juvenile literature.   I. LearningExpress (Organization)
II. Title: Geometry in fifteen minutes a day.
  QA445.5.G46 2011
  516—dc22

                            2010045263

Printed in the United States of America

9 8 7 6 5 4 3 2 1

First Edition

ISBN 978-1-57685-766-3

For more information or to place an order, contact LearningExpress at:
  2 Rector Street
  26th Floor
  New York, NY 10006

Or visit us at:
  www.learnatest.com

ⒸⓄⓃⓉⒺⓃⓉⓈ

CONTRIBUTORS

Kimberly Stafford studied mathematics and education on the verdant grounds of Colgate University, in upstate New York. From there, she went to teach in Japan and Virginia before landing in her current domicile, Los Angeles. In LA, Kimberly spent some time teaching middle school mathematics before beginning her own private tutoring business in 2002. She has since enjoyed the individualized work she does with her students in a one-on-one setting, along with the benefits of having her sunny Southern California days to enjoy outdoors.

Stephen A. Reiss, MBA, is the founder and owner of the Math Magician and the Reiss SAT Seminars, test preparation centers in San Diego. Reiss has authored and coauthored several test preparation books and is a member of MENSA.

INTRODUCTION

DUE TO ITS endless real-world applications, geometry is a particularly useful branch of mathematics. Geometry is something that people find themselves using when they plan a building project at home, buy paint to give their office a new look, or select gardening supplies in the springtime. Having an understanding of geometric concepts can help you solve a wide variety of problems and make smart decisions. We challenge you to begin looking for places where you can put the geometry concepts in this book to use in your everyday life. Finding real-world applications will help you deepen your knowledge of how mathematical principles work together, and it will also strengthen your motivation to fully grasp the material you are studying.

Having a strong foundation in geometry is essential for success in trigonometry and pre-calculus. This book will be a great resource as you try to gain a comprehensive understanding of the concepts covered in a first-year geometry course.

We have sequenced the book so that each lesson builds on the previous lessons. Therefore, it is a good idea to move forward through the book, making sure you fully understand each lesson's material before proceeding to the next lesson. Each lesson has examples, illustrations, and boxed theorems and postulates, along with two to three sets of practice problems to master the presented material. By investing 15 minutes of focused study time per day on each lesson, you should be able to move through this collection of 30 lessons within a month, and as a result, feel confident in your understanding of geometry.

P R E T E S T

THIS PRETEST WILL test your knowledge of basic geometry. The 30 questions are presented in the same order that the topics they cover are presented in this book. The book builds skills from lesson to lesson, so skills you will learn in earlier lessons may be required to solve problems found in later lessons. This means that you may find the problems toward the beginning of the test easier to solve than the problems toward the end of the test.

For each of the 30 multiple-choice questions, circle the correct answer. The pretest will show you which geometry topics are your strengths and which topics you might need to review—or learn for the first time. After taking the pretest, you might realize that you need only to review a few lessons. Or you might benefit from working through the book from start to finish.

PRETEST

1. In the first figure, all of the following are acceptable ways to name the shaded angle EXCEPT

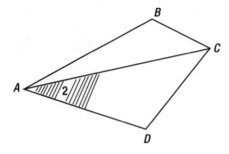

   a. $\angle A$
   b. $\angle 2$
   c. $\angle CAD$
   d. $\angle DAC$

2. In the next figure, D bisects $\overline{NA}$ and N bisects $\overline{IA}$. If $\overline{DA} = 2x - 3$ units, what is the length of $\overline{IA}$?

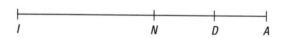

   a. $4x - 6$
   b. $6x - 9$
   c. $8x - 12$
   d. $4x + 9$

3. $\angle K$ is complementary to $\angle M$. If $m\angle M = (2x)°$, then what is the measure of $\angle K$ in terms of x?
   a. $90° + (2x)°$
   b. $90° - (2x)°$
   c. $800° + (2x)°$
   d. $800° - (2x)°$

4. Solve for x.

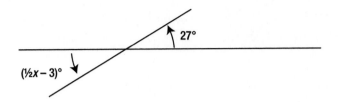

(½x – 3)°   27°

    a. 27
    b. 30
    c. 54
    d. 60

5. In the next figure, $l \parallel k$. $\angle A$ and $\angle B$ are what type of angles?

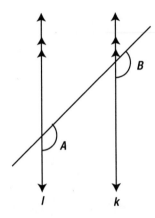

    a. Corresponding angles
    b. Complementary angles
    c. Alternate exterior angles
    d. Same-side interior angles

6. What is the value of x, given that $m \perp n$?

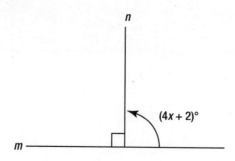

   a. 90

   b. 45

   c. 22

   d. 18

7. In $\triangle CAT$, $\overline{CA} < \overline{AT} < \overline{TC}$. If $\overline{CA} = 12$ and $\overline{TC} = 30$, then what could be a possible length for side $\overline{AT}$?

   a. 17

   b. 18

   c. 19

   d. Cannot be determined.

8. Which of the following statements is false?

   a. An isosceles triangle can also be a right triangle.

   b. The angles of an equilateral triangle will *always* equal 60°.

   c. An isosceles triangle has two congruent angles.

   d. A right triangle has two right angles.

9. In an isosceles triangle $\triangle LEX$, $\overline{LE} \cong \overline{LX}$. If $m\angle LEX = 27°$ find the $m\angle ELX$.

   a. 126°

   b. 153°

   c. 63°

   d. 54°

10. Which expression is equal to the area of the given triangle?

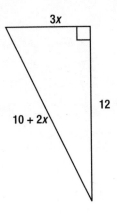

a. $5x + 22$

b. $36x$

c. $120x + 24$

d. $30x + 6x^2$

11. In a right triangle, the lengths of the two legs are 8 cm and 15 cm. Find the length of the hypotenuse.

a. 23 cm

b. 19 cm

c. 7 cm

d. 17 cm

12. $\triangle PIN$ is a 30°-60°-90° triangle. If $\overline{PI}$ measures 14 m, as shown in the accompanying figure, what will be the length of $\overline{PN}$?

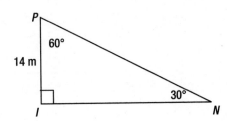

a. 28 m

b. 21 m

c. $14\sqrt{2}$ m

d. $14\sqrt{3}$ m

13. Which postulate would be used to prove that the two triangles below are congruent?

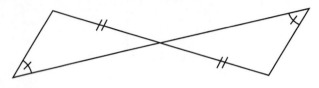

    a. angle–side–angle
    b. side–angle–side
    c. angle–angle–side
    d. side–side–angle

14. What is the measure of an interior angle of a regular hexagon?
    a. 60°
    b. 120°
    c. 72°
    d. 45°

15. Quadrilateral *TOYS* has the following angle measurements: $m\angle T = (y)°$, $m\angle O = (2y)°$, $m\angle Y = (4y)°$, $m\angle Y = (5y)°$. What is the measure of the largest angle in quadrilateral *TOYS*?
    a. 30°
    b. 130°
    c. 150°
    d. 160°

16. Each of the following is a type of parallelogram EXCEPT:
    a. Square
    b. Rectangle
    c. Rhombus
    d. Trapezoid

17. Solve for g: $\frac{15}{24} = \frac{g}{16}$
    a. 10
    b. 12
    c. 11
    d. 13

18. △*ABC* ≈ △*XYZ*. Find the length of side $\overline{XZ}$.

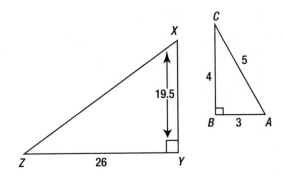

    a. 28
    b. 29.5
    c. 31
    d. 32.5

19. What is the perimeter of the right polygon that follows?

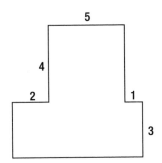

    a. 15 units
    b. 20 units
    c. 26 units
    d. 30 units

20. What is the area of a square that has a perimeter of 24 cm?
    a. 16 cm^2
    b. 24 cm^2
    c. 36 cm^2
    d. 48 cm^2

21. In the figure below, $m\,\widehat{IHF} = 216°$. Find the $m\angle J$.

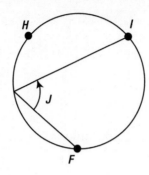

    a. 54°
    b. 72°
    c. 108°
    d. 144°

22. What is the circumference, in terms of π, of a circle with a radius of 6 inches?
    a. 6π in.
    b. 36π in.
    c. 3π in.
    d. 12π in.

23. What is the area, in terms of π, of the shaded region of circle H if $m\angle VHL = 90°$ and $\overline{HL} = 8$ cm?

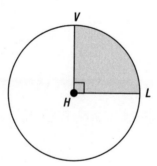

    a. 16π cm²
    b. 64π cm²
    c. 32π cm²
    d. 4π cm²

24. Nina purchases 64 feet of fencing to protect her organic garden project from deer. If she plans on making a rectangular garden with a width of 8 feet, how many feet long can her garden be?

   a. 8 ft.

   b. 16 ft.

   c. 20 ft.

   d. 24 ft.

25. What is the surface area of a rectangular prism that is 8 units long, 6 units wide, and 5 units tall?

   a. 228 units^2

   b. 236 units^2

   c. 240 units^2

   d. 264 units^2

26. What is the volume of the cylinder in the figure? Round your answer to the nearest tenth.

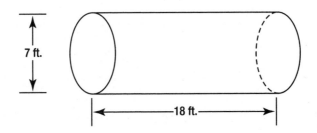

   a. 395.6 ft.^{2}

   b. 791.3 ft.^{2}

   c. 692.4 ft.^{2}

   d. 2,769.5 ft.^{2}

27. What is the midpoint between points $R(-6,-5)$ and $H(4,-3)$?

   a. (-5,1)

   b. (-1,-4)

   c. (-1,1)

   d. (-2,-4)

28. A vertical line has a slope that is
  a. Undefined
  b. Positive
  c. Zero
  d. Negative

29. Which equation is *not* linear?
  a. $5x + 4 = \frac{1}{2}y$
  b. $3x = -12 - 4y$
  c. $\frac{8x}{y} = 12$
  d. $-5 = 20xy$

30. What transformation could be used to create image point $D'(3,-4)$ from preimage point $D(-3,-4)$?
  a. A clockwise rotation of $90°$.
  b. A reflection over the x-axis.
  c. A translation using $T_{-1,0}$
  d. A reflection over the y-axis.

ANSWERS

For any questions that you may have missed, take a look at the lesson listed.

1. a. Lesson 1
2. c. Lesson 2
3. b. Lesson 3
4. d. Lesson 4
5. a. Lesson 5
6. c. Lesson 6
7. c. Lesson 7
8. d. Lesson 8
9. a. Lesson 9
10. b. Lesson 10
11. d. Lesson 11
12. a. Lesson 12
13. c. Lesson 13
14. b. Lesson 14
15. c. Lesson 15
16. d. Lesson 16
17. a. Lesson 17
18. d. Lesson 18
19. d. Lesson 19
20. c. Lesson 20
21. b. Lesson 21
22. d. Lesson 22
23. a. Lesson 23
24. d. Lesson 24
25. b. Lesson 25
26. c. Lesson 26
27. b. Lesson 27
28. a. Lesson 28
29. d. Lesson 29
30. b. Lesson 30

basic geometry concepts

Without geometry life is pointless.
—ANONYMOUS

In this lesson you will become acquainted with the basic definitions of the building blocks of geometry such as points, lines, line segments, rays, planes, and angles.

POINTS, LINES, RAYS, AND PLANES

A **point** is the most basic geometric unit. It looks like a dot that you would make with your pen and is named by a single capital letter. Although points have no direction, length, or size, they are essential in helping you name lines, rays, planes, and angles. Point *A* is demonstrated next.

. *A*

A **line** is a straight collection of points that extend infinitely in both directions, like the horizon over the ocean. Even though it takes only two points to define a line, lines contain an infinite amount of points, which means more points than you can even count. Points that are on the same line are called **collinear**. A line is named by writing any of the two points on it and then placing a double-ended arrow above the letters. So, if you saw the symbol $\overleftrightarrow{AB}$, you would read it "line AB." If you saw this symbol, $\overleftrightarrow{BA}$, you would read it "line BA." The order that you write the letters in does not matter:

$\overleftrightarrow{AB}$ and $\overleftrightarrow{BA}$ both define the line.

A **line** can also be named by writing a single lowercase letter that has been written beside it:

l can be used to define this line.

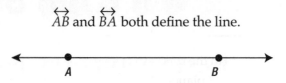

A **line segment** is like a smaller piece of a line. It does not continue endlessly in both directions. Instead, it is a fixed length that has a beginning point and an endpoint. A ruler is a good example of many line segments. (Line segments still contain a countless number of points.) A line segment is named using the letter endpoints with an arrowless line segment above them.

$\overline{CD}$ defines this line segment.

A **ray** is a mix between a line and a line segment. Rays contain one fixed endpoint with the other end of the ray continuing forever. Think of a ray as a shining flashlight. A ray is always named by the endpoint and any other point on the ray, with a one-sided arrow written above.

$\overrightarrow{EF}$ and $\overrightarrow{EG}$ both define this ray.

A **plane** is a collection of points that makes a flat surface, which extends continuously in all directions, like an endless wall. Do you remember that a line is created with just two points, and that points that are on the same line are considered to be **collinear**? When points do NOT lie on the same line, they are **non-collinear**. A plane needs a minimum of three non-collinear points to be defined. In geometry, a plane is usually drawn as a four-sided shape with a capital letter in one of its corners, as in plane *Q* following. Less commonly, planes can be referred to as any three points contained on them, as with plane *XYZ* following.

 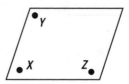

Review the concepts in the following table and then complete the practice questions on lines, rays, and planes.

THE BASIC BUILDING BLOCKS OF GEOMETRY

Figure	Name	Symbol	Read As	Properties	Examples
· A	Point	· A	point A	• has no size • has no dimension • indicates a definite location • named with an italicized uppercase letter	• pencil point • corner of a room
A B ↔ l ↔	Line	$\overleftrightarrow{AB}$ or $\overleftrightarrow{BA}$	line AB or BA or line l	• is straight • has no thickness • an infinite set of points that extends in opposite directions • one dimension	• highway without boundaries • hallway without bounds
A B →	Ray	$\overrightarrow{AB}$	ray AB (endpoint always first)	• is part of a line • has only one endpoint • an infinite set of points that extends in one direction • one dimension	• flashlight • laser beam
A B	Line segment	AB or BA	segment AB or BA	• is part of a line • has two endpoints • an infinite set of points • one dimension	• edge of a ruler • base board
	Plane	None	plane ABC or plane X	• is a flat surface • has no thickness • an infinite set of points that extends in all directions • two dimensions	• floor without boundaries • surface of a football field without boundaries

PRACTICE 1

1. What are the different ways to name the line segments in the accompanying figure?

2. In the figure from question **1**, what are the different ways to name the ray?

3. What are three ways to name the following line?

4. How many points does it take to create a plane?

5. How many points does a line have?

6. True or false: When you're naming a ray, it does not matter which letter comes first.

7. A hose spraying water is most similar to a line, a line segment, or a ray?

ANGLES

An **angle** is created when two rays come together at one common endpoint. This endpoint is called the **vertex** of the angle. The angle is the space in between the two rays, or sides, and is measured in degrees. The number of degrees an angle has identifies what type of angle it is. You will learn more about that in the next lesson. Either the $\angle$ symbol or the $\measuredangle$ symbol is always used when we are discussing angles. An angle can be named in three different ways including:

1. By its vertex (if there are *only* two rays that connect at the vertex).
2. By using three letters, starting with a point on one of the sides, followed by the vertex point, and ending with a point on the remaining side. (The order does not matter as long as the vertex is in the center.)
3. By a number drawn inside the angle to identify it.

The following figure shows the three ways to name an angle.

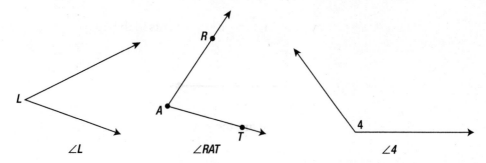

∠L   ∠RAT   ∠4

PRACTICE 2

1. What are three different ways that the following angle can be named?

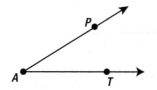

2. An angle is formed when two _____ come together at a common endpoint.

3. How many different angles in the following figure have ray $\overrightarrow{XB}$ as one of the sides? Name them.

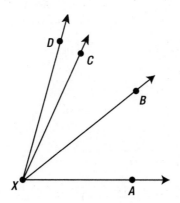

4. Point A in the following figure is called the _____ of the angle.

5. Why is it that "*W*" in the following figure cannot be used to name the angle?

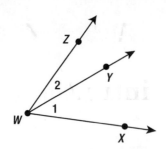

6. Using the figure above, list five ways to name the given angles.

7. True or false: An angle can have more than two vertices.

ANSWERS

Practice 1

1. The line segments can be named $\overline{CA}$, $\overline{AC}$, $\overline{TA}$, or $\overline{AT}$ since the order of the letters does not matter.

2. The ray can only be named $\overrightarrow{CA}$ or $\overrightarrow{CT}$ since the endpoint, *C*, must come first.

3. $\overleftrightarrow{XY}$, $\overleftrightarrow{YX}$, or *m*

4. It takes at least three non-collinear points to make a plane.

5. A line has an infinite, or countless, number of points.

6. False: When you are naming rays, the endpoint must always come first.

7. A hose spraying water is most similar to a ray since there is a definite endpoint.

Practice 2

1. ∠*PAT*, ∠*TAP*, ∠*A*

2. rays

3. Three angles have $\overrightarrow{XB}$ as a side: ∠*AXB*, ∠*BXC*, and ∠*BXD*.

4. vertex

5. "*W*" cannot be used to name either of the angles because more than two rays come together at this vertex.

6. ∠*XWY* or ∠1, ∠*YWZ* or ∠2, and ∠*XWZ*

7. False. Angles have only one vertex, where the two sides, or rays, come together.

2

introduction to angles
and segments

Where there is matter, there is geometry.

—JOHANNES KEPLER

In this lesson you will learn how to use a protractor to measure and create angles. You will also learn the basics of measuring line segments.

USING A PROTRACTOR TO MEASURE ANGLES

A **protractor** is a special ruler that is used to measure angles. Unlike distances, which can be measured in various units like inches, feet, or miles, angles are measured only in **degrees**. Degrees are always notated with the symbol °. The expression "$m\angle B = 40°$" is read, *The measure of an angle B is 40 degrees.*

Most protractors have two scales along their arc. The lower scale, starting with zero on the right, is used to **measure angles** that open in a counterclockwise arc. To measure a counterclockwise angle, place the vertex in the circular opening at the base of the protractor and extend the bottom ray through the 0° mark out to the right. In the following figure, the counterclockwise angle drawn is 15°.

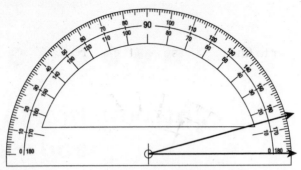

The second scale along the top, which begins with zero on the left, measures angles that open in a clockwise arc. To measure a clockwise angle, place the vertex in the circular opening at the base of the protractor and extend the bottom ray through the 0° mark out to the left. In the following figure, the clockwise angle drawn is 120°.

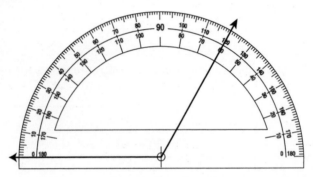

Before we practice measuring angles, there is one more important concept to discuss. Sometimes, you will need to measure an angle that does not pass through 0°, but instead passes through two different measurements. When this happens, subtract the smaller measurement from the larger measurement in order to identify the number of degrees in the angle. For example, say that in the following figure, you need to find the $m\angle BXC$. $\overrightarrow{XB}$ goes through 30° and $\overrightarrow{XC}$ goes through 110°, so in order to find $m\angle BXC$, subtract 30° from 110° to get 80°. Therefore, $m\angle BXC = 80°$.

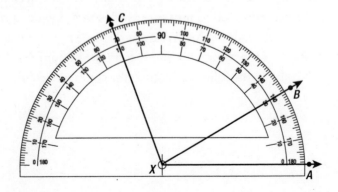

PRACTICE 1

Use the protractor to find the measure of the given angles.

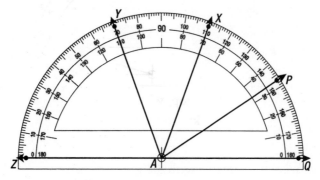

1. $m\angle QAP =$

2. $m\angle PAX =$

3. $m\angle QAY =$

4. $m\angle ZAX =$

5. $m\angle ZAP =$

6. $m\angle PAY =$

7. $m\angle QAZ =$

USING A PROTRACTOR TO DRAW ANGLES

Using a protractor to draw angles is easy once you understand how to measure angles. First, draw a ray and place the protractor's circular opening over the point that will be the angle's vertex. Then align that ray with the 0° mark. Find the degree of the angle you are drawing, and make a dot that marks the endpoint of the second ray. Last, remove the protractor and connect the vertex with your drawn endpoint. In the following figure, we are ready to make a 75° angle.

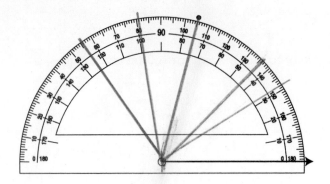

PRACTICE 2

Use a protractor to draw angles with the given measures.

1. 45°

2. 75°

3. 100°

4. 125°

5. 32°

MEASURING LINE SEGMENTS

Line segments are easy to measure when you have a ruler, but often in geometry you will be asked to determine the length of a missing segment with only limited information. In order to indicate the length of a line segment, two letters defining the segment are written with a small line over them. For example, $\overline{BC}$ can be read, *The length of line segment* $\overline{BC}$.

As long as points are all on the same line, the sum of all of the non-overlapping segments will equal the length of the entire line segment. For example, notice that in the following figure, $\overline{AB} = 5$ and $\overline{CD} = 6$, but that segment $\overline{BC}$ has no given measurement. If you were given that $\overline{AD} = 18$, you could determine that $\overline{BC}$ measures 7 units, by subtracting 5 and 6 from 18.

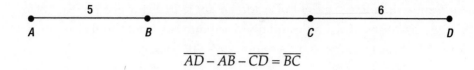

$$\overline{AD} - \overline{AB} - \overline{CD} = \overline{BC}$$

Therefore since $18 - 5 - 6 = 7$, $\overline{BC} = 7$ units.

Determining segment length gets more challenging when some of the given segments are overlapping. For example, consider this figure.

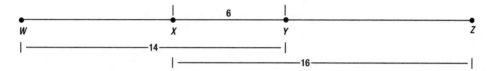

If you were given that $\overline{WY} = 14$, $\overline{XY} = 6$, and $\overline{XZ} = 16$, how would you determine the length of $\overline{WZ}$? You cannot add all three segments together, since $\overline{XY}$ would be included three times, since it is part of both $\overline{WY}$ and $\overline{XZ}$. Therefore, the correct way to do this is to add $\overline{WY}$ and $\overline{XZ}$, and **then subtract** $\overline{XY}$. This would look like: $(14 + 16) - (6) = 24$ units.

MIDPOINTS AND BISECTORS

Line segments can have both **midpoints** and **bisectors**. A **midpoint** is a point that divides a line segment into two segments of equal length. A **bisector** is a line, ray, or segment that goes through a line segment's midpoint, dividing the segment into two equal parts. In the next figure, point I is the midpoint of $\overline{JM}$. In the same illustration, $\overline{BN}$ is the bisector of $\overline{JM}$ since it passes through midpoint I.

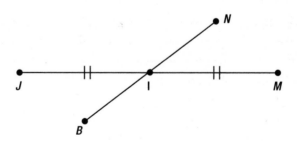

Similar to a line segment bisector, an **angle bisector** is a ray that passes through the vertex of an angle and divides it into two equal angles. In the next figure, $\overrightarrow{YW}$ bisects $\angle XYZ$ and therefore $m\angle XYW = m\angle WYZ$.

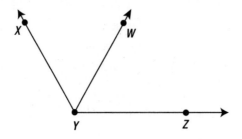

PRACTICE 3

Use the following figure to answer questions 1–5. Note: The figure is NOT drawn to scale.

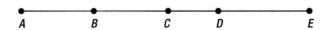

1. If $\overline{AD} = 8$, $\overline{CE} = 6$, and $\overline{CD} = 2$, what is the length of $\overline{AE}$?

2. If $\overline{AB} = 13$, B is the midpoint of $\overline{AC}$, and C is the midpoint of $\overline{AE}$, what is the length of $\overline{AE}$?

3. If $\overline{AB} = 3.4$, $\overline{BD} = 7.6$, and $\overline{AE} = 14.8$, what is the length of $\overline{DE}$?

4. If B is the midpoint of $\overline{AC}$, and C is the midpoint of $\overline{AE}$, what is the length of $\overline{BC}$ if $\overline{AE} = 56$?

5. If $\overline{AD}$ is twice as large as $\overline{DE}$, and $\overline{AE} = 36$, what is the length of $\overline{AD}$?

ANSWERS

Practice 1

1. $33°$

2. $37°$ ($70° - 33° = 37°$)

3. $130°$

4. 110°

5. 147°

6. 97° (130° − 33° = 97°)

7. 180°

Practice 2

1.

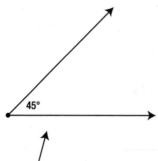

2.

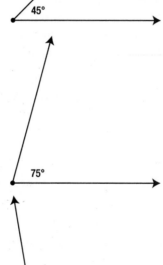

3.

4.

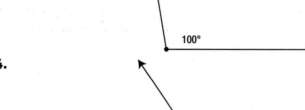

5.

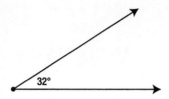

32°

Practice 3

1. $\overline{AE} = 12$. Add $\overline{AD} = 8$ to $\overline{CE} = 6$ and subtract $\overline{CD} = 2$, since $\overline{CD}$ is counted twice where the two segments overlap.

2. $\overline{AE} = 52$. Since $\overline{AB} = 13$ and B is the midpoint of $\overline{AC}$, then $\overline{AC} = 26$. Since C is the midpoint of $\overline{AE}$, then $\overline{AE} = 52$.

3. $\overline{DE} = 3.8$. Subtract the smaller segments from the whole: $14.8 - 7.6 - 3.4 = 3.8$.

4. $\overline{BC} = 14$. Since $\overline{AE} = 56$ and C is its midpoint, $\overline{AC} = 28$. Since B is the midpoint of $\overline{AC}$, then $\overline{BC} = 14$.

5. $\overline{AD} = 24$. Since $\overline{AD}$ is twice as large as $\overline{DE}$, let $\overline{AD} = 2x$ and $\overline{DE} = x$ and then write the formula, $2x + x = 36$. Solving for x you see that $x = 12$, so $\overline{AD}$ is twice that.

3

angles in the plane

Geometry is just plane fun.
—Anonymous

In this lesson you will learn about the different types of angles and about complementary and supplementary angle relationships.

DEFINING ANGLES

Angles are defined by their relationship to the degree measures of 90° and 180°. However, in geometry, drawings involving angles are not drawn to perfect scale. You should usually be able to tell which of the following four categories an angle falls into.

An angle that is larger than 0° but smaller than 90° is called an **acute angle**. (A clever way to remember this is to think, "Oh, isn't that small angle so *cute!*")

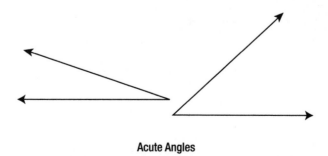

Acute Angles

Right Angles

An angle that measures 90° is called a **right angle**. Since right angles are extra special angles, they have their own symbol to show they are 90°. A small square is drawn at the vertex of right angles to note their unique nature, as shown in the following figure. We'll learn about the special properties of right angles later in this and in the following lesson. (It might help you to remember that a right angle has a side that rises directly up from its base by thinking, "Oh, isn't that angle so up*right*!")

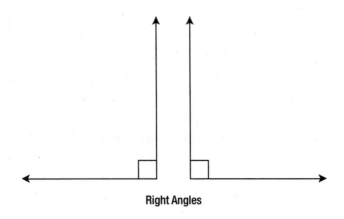

Right Angles

Obtuse Angles

An angle that is larger than 90° but smaller than 180° is called an **obtuse angle**. (Do you like rhymes? "That *obtuse* is as big as a *moose*!")

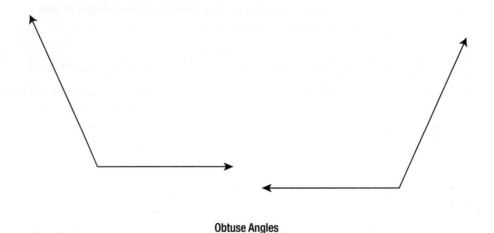

Obtuse Angles

Straight Angles

An angle that is 180° looks like a straight line and is called a **straight angle**. Take note that there are no names for angles larger than 180°.

Straight Angle

PRACTICE 1

Answer the following questions about angle classifications.

1. If a straight angle is bisected, the two angles that are created are

_____.

2. If a straight angle is broken into three angles of equal measure, then each of those angles is _____.

3. True or false: An acute angle plus an acute angle will always give you an obtuse angle.

4. A straight angle minus an obtuse angle will give you an acute angle. Is this statement sometimes, always, or never true?

5. A right angle plus an acute angle will give you an obtuse angle. Is this statement sometimes, always, or never true?

6. An obtuse angle minus an acute angle will give you an acute angle. Is this statement sometimes, always, or never true?

7. Which of the following angles is not acute, obtuse, right, or straight?
   a. 180°
   b. 167°
   c. 195°
   d. 3.7°

Use the following images to complete questions 8–11.

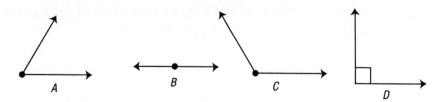

8. Which angle is acute?

9. Which angle is a right angle?

10. Which angle is a straight angle?

11. Which angle is obtuse?

SUPPLEMENTARY AND COMPLEMENTARY ANGLES

Do you remember reading earlier that angles are defined by their relationship to 90° and 180°? Now you know how to classify angles based on their specific degree measurement. Next, you will learn about the special relationships that can exist between two or more angles. When two angles have measurements that sum to 90°, we say that the angles are **complementary**. The next figure shows two pairs of complementary angles. Notice that complementary angles can be two completely separate angles, as with ∠1 and ∠2. Also, complementary angles can share a common side and form a right angle together, as with ∠3 and ∠4. Whenever acute angles form a right angle, they are complementary.

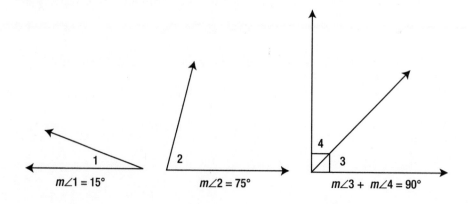

When two angles have measurements that sum to 180°, we say that the angles are **supplementary**. The next figure shows two pairs of supplementary angles. Supplementary angles can be two completely separate angles, as with ∠5 and ∠6. Supplementary angles can also share a common side and form a straight angle, as is the case with ∠7 and ∠8. When supplementary angles are adjacent and form a straight angle, they are called a **linear pair**.

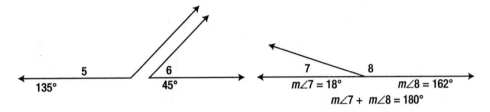

The following figure illustrates how three or more angles can be complementary or supplementary.

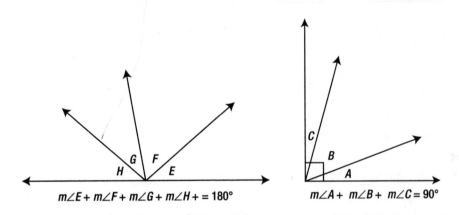

PRACTICE 2

1.

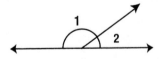

Are angles 1 and 2 supplementary, complementary, or neither?

For questions **2–5**, find the measure of the complement and the supplement for each of the given angles.

2. $\angle 2 = 15°$: complement = _____ ; supplement = _____

3. $\angle 3 = 78.5°$: complement = _____ ; supplement = _____

4. $\angle 4 = 137°$: complement = _____ ; supplement = _____

5. $\angle 5 = 180°$: complement = _____ ; supplement = _____

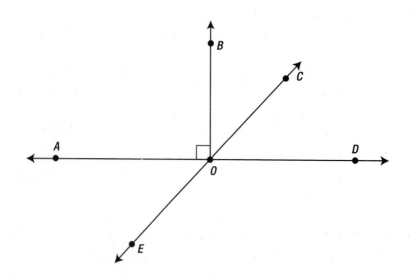

6. Name two supplementary angles of $\angle DOE$.

7. Name a pair of complementary angles.

For questions 8–12, state whether the following statements are true or false.

8. Complementary angles must be acute.

9. Supplementary angles must be obtuse.

10. Two acute angles can be supplementary.

11. Any two right angles are supplementary.

12. Two acute angles are always complementary.

ANSWERS

Practice 1

1. right angles $\frac{180°}{2} = 90°$
2. an acute angle $\frac{180°}{3} = 60°$
3. False. Two small acute angles do not have to have a sum greater than 90°.
4. Always. Such as: $180 - 102 = 78$
5. Always
6. Sometimes
7. c. Remember, an obtuse angle must be *between* 90° and 180°.
8. a.
9. d.
10. b.
11. c.

Practice 2

1. supplementary
2. complement = 75°; supplement = 165°
3. complement = 11.5°; supplement = 101.5°
4. complement = does not exist; supplement = 43°
5. 180° is a straight angle and it does not have a complement or a supplement.
6. $\angle AOE$ and $\angle COD$
7. $\angle BOC$ and $\angle COD$
8. True
9. False—If there are only two angles, there are two cases: Either one angle must be obtuse and the other angle must be acute *or* both angles must be right angles.
10. False
11. True
12. False—Acute angles *can* be complementary, but they do not *have* to be.

special pairs of angles

Geometry is the foundation of all painting.
—ALBRECHT DÜRER

In this lesson you will first learn how to identify adjacent and vertical angles, and then you will gain an understanding of the relationships between these types of angles.

ADJACENT ANGLES

Adjacent angles are angles that share a common vertex, have one common side, and have no interior points in common. In the next figure, $\angle ZWY$ and $\angle YWX$ are adjacent since they share side WY. Since they are not overlapping, they also have no common interior points.

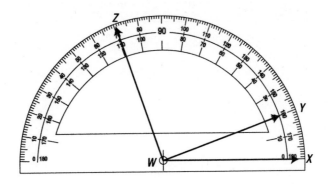

It is important to understand that in this figure, $\angle ZWY$ and $\angle ZWX$ are NOT adjacent since the interior of $\angle ZWY$ is included in $\angle ZWX$. Remember, adjacent angles do not have any interior points in common. The word adjacent means *next to*, so adjacent angles are next to each other, but cannot overlap with one another.

Just as with numbers and variables, addition and subtraction can be used to combine or separate measures of angles. Addition and subtraction are especially helpful when you are working with adjacent angles. For example, in this figure, $m\angle 1 + m\angle 2 = m\angle ZWX$.

Another way to look at this could be to use subtraction in the following way: $m\angle ZWX - m\angle 1 = m\angle 2$.

TIP: The sum of two adjacent angles equals the measure of the larger angle formed by their non-common sides.

You should remember from the last lesson that two angles are supplementary when they sum to 180°. When two adjacent angles are supplementary, they are referred to as a **linear pair** and they form a straight angle, as in the following.

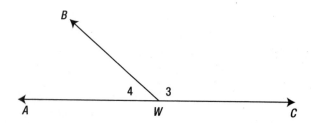

In this figure, subtraction can be used to form the following relationships:

$$180° - m\angle 3 = m\angle 4$$

$$180° - m\angle 4 = m\angle 3$$

PRACTICE 1

Use this figure and the given angle measurements to answer questions **1–5**.

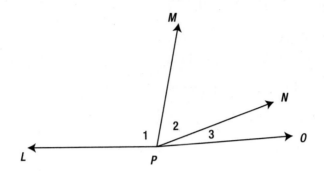

1. $m\angle 1 = 93°$
   $m\angle 2 = 48°$
   Find $m\angle LPN$.

2. $m\angle MPO = 78°$
   $m\angle 2 = 42°$
   Find $m\angle 3$.

3. $m\angle 1 = 3x$
   $m\angle 2 = 2x$
   $m\angle 3 = x$
   Find $m\angle LPO$ in terms of x.

4. $m\angle LPO = 175°$
   $m\angle 3 = 33°$
   $m\angle 1 = 95°$
   Find $m\angle 2$.

5. $m\angle 1 = (4y + 15)°$
   $m\angle 2 = (3y - 6)°$
   $m\angle LPN = 149°$
   Find $m\angle 1$ and $m\angle 2$.

Use the following figure and the given angle measurements to answer questions 6–11.

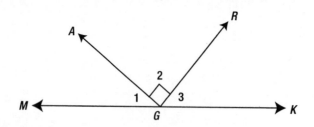

6. $m\angle 1 = 37.5°$
Find $m\angle MGR$.

7. $m\angle 3 = (2x + 10)°$
$m\angle AGK = 144°$
Find the value of x.

8. If $m\angle 1$ is one-half the $m\angle 3$, find the measure of angles 1 and 2. (*Remember:* $\angle AGR$ is a right angle and $\angle MGK$ is a straight angle.)

9. True or false: Angles 1 and 3 are adjacent angles.

10. Definition: Angles 1, 2, and 3 are three adjacent angles that form a
_____ angle.

11. $m\angle 1 = (3x + 3)°$
$m\angle 3 = (6x - 3)°$
Find $m\angle 1$, $m\angle 3$, and $m\angle AGK$.

VERTICAL ANGLES

When two rays, line segments, or lines intersect, the pair of non-adjacent angles that do not have common sides are called **vertical angles**. The intersection of any two rays forms two pairs of **congruent** vertical angles. When two angles are congruent it means they have the same angle measurement. Vertical angles are sometimes also called **opposite angles**, since they are opposite each other, rather than adjacent.

TIP: When two rays, line segments, or lines intersect, two pairs of vertical angles are formed.

In the following figure, angles 1 and 3 are one pair of congruent vertical angles. Angles 2 and 4 are another pair of congruent vertical angles. Notice how vertical angles are not adjacent, since they do not share a common side, but instead, they are opposite each other.

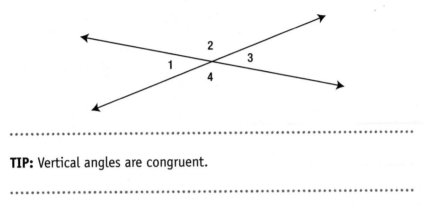

TIP: Vertical angles are congruent.

Most of the time, intersecting lines will form one pair of obtuse vertical angles and one pair of acute vertical angles, as in the preceding figure. Angles 1 and 3 are a pair of congruent acute angles, while angles 2 and 4 are a pair of congruent obtuse angles. In this case, the obtuse vertical angles will be congruent and the acute vertical angles will also be congruent. If two lines intersect to form a 90° angle, then four right angles are formed, as in the following figure.

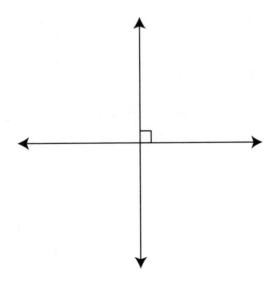

PRACTICE 2

Use this figure to answer questions 1–5.

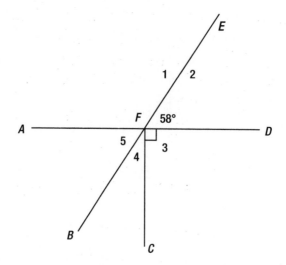

1. Name one pair of vertical angles.

2. Find $m\angle CFE$.

3. Find $m\angle 5$.

4. Find $m\angle 4$.

5. True or false: Angles 1 and 4 are vertical angles.

State whether the following statements are true or false.

6. A pair of vertical angles can be complementary.

7. A pair of vertical angles can be supplementary.

8. Vertical angles are always acute.

Use this figure to answer questions **9–12**.

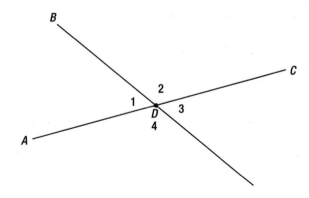

In the figure, $m\angle 1 = 72°$, $m\angle 2 = (2x + 6)°$.

9. Find $m\angle 3$.

10. Find $m\angle 2$.

11. Find $m\angle ADC$

12. Find the value of x.

ANSWERS

Practice 1

1. $m\angle LPN = 141°$
2. $m\angle 3 = 36°$
3. $m\angle LPO = 6x$
4. $m\angle 2 = 47°$
5. $m\angle 1 = 95°$ and $m\angle 2 = 54°$
6. $m\angle MGR = 127.5°$
7. $x = 22$
8. $m\angle 1 = 30°$ and $m\angle 3 = 60°$
9. False
10. Straight
11. $m\angle 1 = 33°$, $m\angle 3 = 57°$, and $m\angle AGK = 147°$

Practice 2

1. Angles 5 and 2 are vertical angles.
2. $m\angle CFE = 148°$
3. $m\angle 5 = 58°$
4. $m\angle 4 = 32°$
5. False
6. True. Two vertical angles can each equal 45°.
7. True. Two vertical angles can each equal 90°.
8. False
9. $m\angle 3 = 72°$
10. $m\angle 2 = 108°$
11. $m\angle ADC = 180°$
12. $x = 51$

angles formed by parallel lines

If parallel lines meet at infinity—infinity must be a very noisy place with all those lines crashing together.
—Anonymous

In this lesson you will learn about transversals, parallel lines, and the relationships that exist between the angles formed by these lines.

TRANSVERSALS

A **transversal** is a line that passes through two or more lines at two separate points on each line. In the following figure, line *M* is a transversal through lines *J* and *F*, but line *N* is not a transversal, since it crosses lines *J* and *F* at the same point.

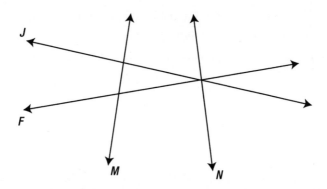

Transversals can sometimes be difficult to recognize, but they are very important since they define specific relationships between the angles they form. Use the following practice questions to become familiar with transversals.

PRACTICE 1

Use the following figure to answer questions 1–4.

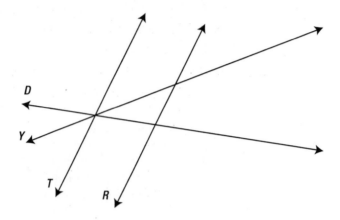

1. Is line D a transversal? Why or why not?

2. Is line Y a transversal? Why or why not?

3. Is line T a transversal? Why or why not?

4. Is line R a transversal? Why or why not?

SKEW AND PARALLEL LINES

If you were to draw a pair of coplanar lines (lines in the same plane), there are three ways to do this. One way is with the two lines intersecting at one and only one point. The point where they cross each other is called the *point of intersection*. These are seen in the following figure.

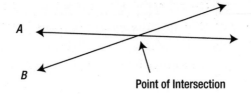

Skew Lines

A second way to draw two lines is to construct them on top of one another. In this case, they share all of their points in common, and are essentially the same line. The third way to draw two lines is so that they would never intersect. In this case, they are called **parallel** lines if they are in the same plane, but if they are in different planes they are considered to be **skew** lines. In order to indicate that two lines are parallel in an illustration, they are drawn with single-arrow or double-arrow markings on them, which you can see in the following figure. In writing, the symbol $\parallel$ is used to indicate parallel lines. In order to write that line K is parallel to line S, the notation $K \parallel S$ is used.

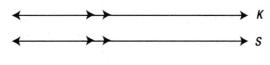

Parallel Lines

TRANSVERALS THROUGH PARALLEL LINES

When a transversal intersects two parallel lines, special angle relationships are formed. In the next figure, lines P and Q are parallel and line X is a transversal that forms eight numbered angles.

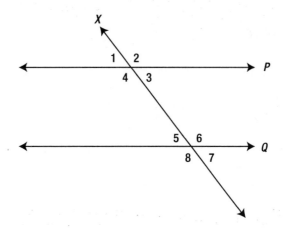

You can probably tell that all of the odd-numbered angles are acute angles: ∠1, ∠3, ∠5, and ∠7. All of the even-numbered angles are obtuse angles: ∠2, ∠4, ∠6, and ∠8. When a transversal intersects parallel lines, all of the acute angles are congruent to one another. Similarly, all of the obtuse angles are congruent to one another.

....

TIP: When a transversal intersects parallel lines, all of the acute angles formed are congruent and all of the obtuse angles are congruent.

....

SPECIAL ANGLE RELATIONSHIPS

Additionally, there are special names and relationships for the pairs of angles formed by a transversal passing through parallel lines.

Corresponding Angles

Corresponding angles have the same relative position between the parallel line and the transversal. For example, ∠2 and ∠6 are both the top right angle formed by the transversal and the parallel line. Therefore, we would say that ∠2 and ∠6 are *corresponding angles*. (See the shaded angles in this figure.) Corresponding angles are always congruent. Can you find another pair of corresponding angles?

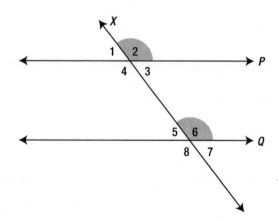

TIP: Corresponding angles are congruent.

Alternate Exterior Angles

Exterior means *outside,* and *alternate* means *other* or *opposite.* **Alternate exterior angles** are the angles *outside* the parallel lines and on the *opposite* sides of the transversal. For example, $\angle 2$ is on the *right* side of transversal X, and is located on the outside of parallel line P. $\angle 8$ is on the *left* side of transversal X, and is located on the outside of parallel line Q. Therefore, $\angle 2$ and $\angle 8$ are *alternate exterior* angles. (See the shaded angles in the following figure.) Alternate exterior angles are always congruent. Can you find another pair of alternate exterior angles?

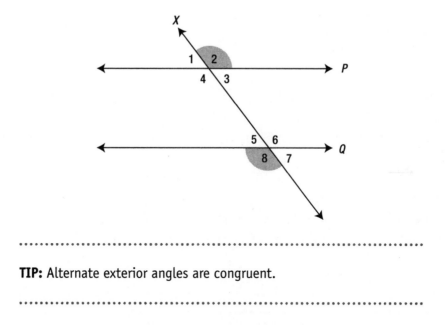

TIP: Alternate exterior angles are congruent.

Alternate Interior Angles

Interior means *inside,* and we already know what *alternate* means. **Alternate interior angles** are the angles *inside* the parallel lines and on *opposite* sides of the transversal. For example, $\angle 3$ is on the *right* side of transversal X, and is located inside the parallel lines P and Q. $\angle 5$ is on the *left* side of transversal X, and is also located on the inside of parallel lines P and Q. Therefore, $\angle 3$ and $\angle 5$ are *alternate interior angles*. (See the shaded angles in the next figure.) Alternate interior angles are congruent. Can you find another pair of alternate interior angles?

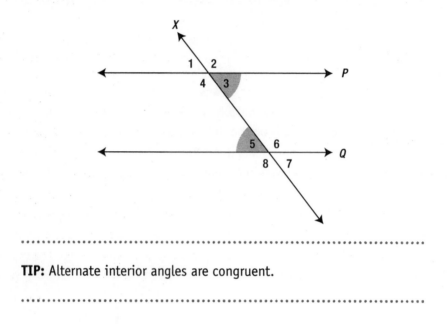

....

TIP: Alternate interior angles are congruent.

....

Same-Side Interior Angles

Same-side interior angles are the angles *inside* the parallel lines, on the *same* side of the transversal. For example, $\angle 3$ and $\angle 6$ are both on the *right* side of transversal X. They are also both located on the inside of parallel lines P and Q. Therefore, $\angle 3$ and $\angle 6$ are *same-side interior angles*. (See the shaded angles in the following figure.) Same-side interior angles are supplementary and add up to 180°. Can you find another pair of same-side interior angles?

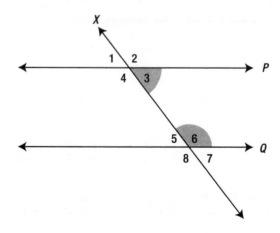

TIP: Same-side interior angles are supplementary.

The property above can be extended to include all pairs of acute and obtuse angles formed with parallel lines and transversals. When two parallel lines are intersected by a transversal, any pair of one acute and one obtuse angle is supplementary.

TIP: When a transversal intersects parallel lines, any pair of one obtuse and one acute angle will be supplementary.

Note: Even when a transversal intersects two non-parallel lines, the same angle relationship names can be used. The properties of congruence will not hold true if the intersected lines are not parallel, but it is still correct to refer to angles as "corresponding," "alternate exterior," and "same-side interior."

PRACTICE 2

Use the non-parallel lines in the next figure to practice identifying relationships between angles in questions **1–6**.

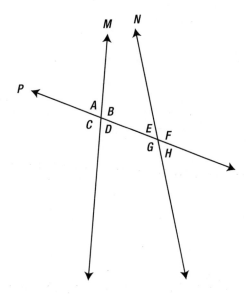

1. Name two pairs of *alternate exterior angles*.

2. Name two pairs of *same-side interior angles*.

3. Name four pairs of *corresponding angles*.

4. Name two pairs of *alternate interior angles*.

5. Complete the following statement and explain why the lines would be parallel: If ∠D were congruent to ∠_____ or to ∠_____, then line M and line N would be parallel.

6. Complete the following statement and explain why the lines would be parallel: If ∠F were congruent to ∠_____ or to ∠_____, then line M and line N would be parallel.

REVIEW OF ANGLE RELATIONSHIPS

Review the facts about special angle relationships in the box and then finish the last set of practice problems.

....

TIP: When a transversal intersects parallel lines:

- Corresponding angles are congruent.
- Alternate interior angles are congruent.
- Alternate exterior angles are congruent.
- Same-side interior angles are supplementary.
- Any pair of acute and obtuse angles are supplementary.

....

PRACTICE 3

Use the parallel lines in the next figure to practice identifying relationships between angles in questions 1–8.

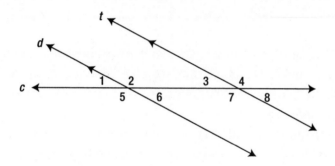

1. Which angles have the same measure as ∠2?

2. ∠1 is an alternate exterior angle to which angle?

3. What is the relationship between ∠6 and ∠7?

4. If $m\angle2 = (4x + 10)°$ and $m\angle3 = (2x – 10)°$, find the measure of ∠2 and ∠3.

5. If the measure of ∠5 is 20 less than four times the measure of ∠8, find the $m\angle5$ and $m\angle8$.

6. If $m\angle7 = (5x + 26)°$ and $m\angle4 = (7x – 18)°$, find $m\angle7$ and $m\angle4$.

7. ∠3 and ∠8 are called _____ angles.

8. ∠4 is supplementary to which four angles?

ANSWERS

Practice 1

1. Line *D* is a transversal through lines *T* and *R* only since it intersects them at two different points.
2. Line *Y* is a transversal through lines *T* and *R* only since it intersects them at two different points.
3. Line *T* is not a transversal since it intersects lines *D* and *Y* at the same point.
4. Line *R* is a transversal through lines *D* and *Y* only since it intersects them at two different points.

Practice 2

1. ($\angle A$ and $\angle H$) and ($\angle C$ and $\angle F$) are two pairs of *alternate exterior angles.*

2. ($\angle B$ and $\angle E$) and ($\angle D$ and $\angle G$) are two pairs of *same-side interior angles.*

3. ($\angle A$ and $\angle E$), ($\angle B$ and $\angle F$), ($\angle C$ and $\angle G$), and ($\angle D$ and $\angle H$) are four pairs of *corresponding angles.*

4. ($\angle D$ and $\angle E$) and ($\angle B$ and $\angle G$) are two pairs of *alternate interior angles.*

5. $\angle E$ (same-side interior angles) or $\angle H$ (corresponding angles)

6. $\angle C$ (alternate exterior angles) or $\angle B$ (corresponding angles)

Practice 3

1. $\angle 5$, $\angle 4$, $\angle 7$

2. $\angle 8$

3. $\angle 6$ and $\angle 7$ are same-side interior angles, and they are supplementary.

4. Same side interior angles are supplementary, so $m\angle 2 + m\angle 3 = 180$:

$(4x + 10) + (2x - 10) = 180$

$6x = 180$, so $x = 30$

$m\angle 2 = (4(30) + 10) = 130$

$m\angle 3 = (2(30) - 10) = 50$

$m\angle 2 = 130°$ and $m\angle 3 = 50°$

5. $m\angle 5 = 140°$ and $m\angle 8 = 40°$

6. Angles 7 and 4 are vertical, which means $m\angle 7 = m\angle 4$:

$(5x + 26)° = (7x - 18)°$

$26 + 18 = 7x - 5x$

$44 = 2x$, so $x = 22$

$(5(22) + 26) = 136$ and $(7(22) - 18) = 136$

$m\angle 7 = m\angle 4 = 136°$

7. vertical (or opposite)

8. $\angle 1$, $\angle 3$, $\angle 6$, and $\angle 8$ are all supplementary to $\angle 4$

perpendicular lines

*Since the mathematicians have invaded the theory of
relativity, I do not understand it myself any more.*
—ALBERT EINSTEIN

In this lesson you will learn about the special properties of perpendicular lines
and perpendicular transversals.

DEFINING PERPENDICULAR LINES

In an earlier lesson you learned about lines that intersect to form adjacent
obtuse and acute angles. Another way that two lines can intersect is to form
right angles, which you probably remember measure 90°. These are very
important angles to understand since there are many real-world applications
of right angles, such as the angle between a street light and the road, the angle
between a wall and the floor of a house, and the angle between two adjoin-
ing pieces of a picture frame. When two lines intersect to form a 90° angle,
those lines are called **perpendicular**. The symbol ⊥ is used to indicate that two
lines are perpendicular. Another way to show that two lines are perpendicular is to draw a small square in the angle where the lines meet, as in the following figure.

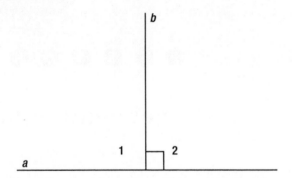

It would be correct here to write, "$a \perp b$." Notice that $\angle 1$ and $\angle 2$ make a straight angle that measures 180°. Although only $\angle 2$ is marked as being 90°, since $\angle 1$ and $\angle 2$ are supplementary, it must be true that $m\angle 1$ is also equal to 90°.

....

TIP: When two adjacent angles form a straight angle, if one of the angles is a right angle, then the second angle is also 90°.

....

When two lines intersect to form four angles, if one of those angles is a right angle, then all four angles are right angles. Observe this in the next figure, where $g \perp h$. In this case, $\angle 1 = \angle 2 = \angle 3 = \angle 4 = 90°$.

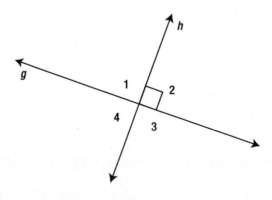

....

TIP: When two intersecting lines are perpendicular, four right angles are formed.

....

In geometry, it is common that figures are not drawn to scale. Therefore, you should never assume additional information that is not given to you. Unless you see one of the two symbols that indicate that an angle is a right angle (either the ⊥ or the square in the angle), you cannot be certain that an angle formed by two seemingly perpendicular lines measures 90°.

PRACTICE 1

Use the figure in the next section to answer the following questions.

1. If $m\angle 2 = (3x - 9)°$, find the value of x.

2. True or false: Angles 3 and 4 are complementary.

3. True or false: You cannot assume that $\angle 3$ is a right angle since it does not have the box drawn in.

4. If $m\angle 1 = (5x + 10)°$ and $m\angle 4 = (8x - 2y)°$, find the values of x and y.

TRANSVERSALS AND PERPENDICULAR LINES

In the previous lesson you learned about the angle relationships that are formed when a transversal intersects two parallel lines. You studied the relationships formed when the transversal creates obtuse and acute angles. A more specific case is when a transversal is perpendicular to a pair of lines as in the next figure.

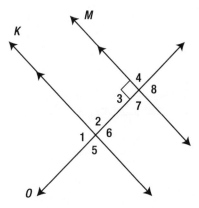

In this case, only $\angle 3$ is marked as 90° with the box at the intersection of the lines. We just learned that when two lines are perpendicular, all four of the angles formed are 90°. Therefore, we know that $\angle 3 = \angle 4 = \angle 7 = \angle 8 = 90°$. Remembering that ∠3 is congruent to its alternate interior angle, ∠6, we can conclude that ∠6 is also a right angle. Once we are certain that ∠6 = 90°, we can conclude that angles 1, 2, and 5 are also right angles. Therefore, when a transversal intersects a pair of parallel lines to form one perpendicular angle, all eight angles formed are right angles.

....

TIP: When a transversal is perpendicular to two parallel lines, it forms eight right angles.

....

PRACTICE 2

Use the following figure to answer questions **1–10**.

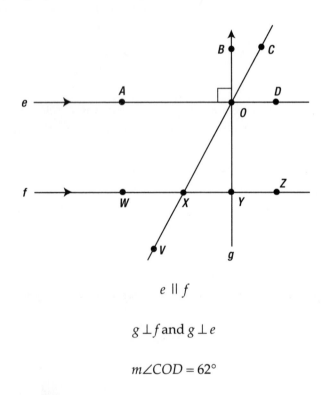

$e \parallel f$

$g \perp f$ and $g \perp e$

$m\angle COD = 62°$

1. Find $m\angle BOC$.

2. Find $m\angle AOC$.

3. Find $m\angle WXO$.

4. Find $m\angle WXV$.

5. Find $m\angle OYZ$.

6. True or false: $\angle WXV$ is an alternate exterior angle to $\angle COD$.

7. True or false: $\angle BOC$ and $\angle AOX$ are vertical angles.

8. True or false: $\angle DOC$ and $\angle XOA$ are vertical angles.

9. If $m\angle AOX = (4x - 26)°$ and $m\angle XOY = (x + 6)°$, find the value of x.

10. $\angle XYO$ is a/an _____ angle.

PERPENDICULAR BISECTORS

The prefix "bi" means two, such as in the word *bi*cycle. The word "bisect" means to cut something into two equal parts. In geometry, when something is bisected, it is cut into two equal or congruent parts. A **perpendicular bisector** is a special type of bisector. It is a line or line segment that cuts a line segment into two equal parts while *also* creating 90° angles. In the next figure, line j is a *perpendicular bisector to both line segments PX, QY, and RZ*. In order to indicate that two line segments are congruent, an equal number of hash marks (‖) are made through line segments.

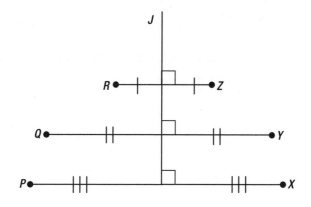

PRACTICE 3

Use the following figure to answer questions **1–4**.

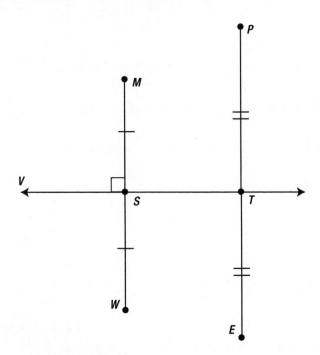

1. The four hash marks on line segments *PT* and *ET* indicate that

_____.

2. Given the markings in this figure, line *V* is the _____ of line segment *MW*.

3. If the length of *MS* is 4*h* + 12 and the length of *WS* is 6*h* – 12, find the length of line segment *MW*.

4. True or false: It is certain that line *V* is a perpendicular bisector of line segment *PE*.

ANSWERS

Practice 1

1. $x = 33$
2. False; they are supplementary.
3. False; since $\angle 2$ and $\angle 3$ form a straight line, and $m\angle 2 = 90°$, then $m\angle 3 = 90°$.
4. $x = 16$ and $y = 19$

Practice 2

1. $m\angle BOC = 28°$
2. $m\angle AOC = 118°$
3. $m\angle WXO = 118°$
4. $m\angle WXV = 62°$
5. $m\angle OYZ = 90°$
6. True
7. False
8. True
9. $x = 22$
10. right / $90°$

Practice 3

1. Line segments PT and ET are congruent.
2. Perpendicular bisector
3. Since $MS = WS$, solve to get $x = 12$. Then the length of MS is $4(12) + 12 = 60$, and WS will be $6(12) - 12 = 60$. The entire length of $MW = 120$.
4. False—Even though line segments PT and ET are equal, there is no way to be sure that line V is a perpendicular bisector of PE.

introduction to triangles

The only angle from which to approach
a problem is the TRY-Angle.
—ANONYMOUS

In this introductory lesson on triangles you will learn about the sum of the interior angles of triangles as well as about acute, obtuse, and right triangles.

SIDES AND ANGLES OF TRIANGLES

A **polygon** is a closed shape that is made up of three or more straight line segments. You will learn more about polygons in future lessons, but the focus here will be on the most basic polygon, the triangle. The most important thing to remember about this three-sided polygon is that the sum of its interior angles is 180°. Remembering that a straight angle has 180°, use the following figure and steps to go through a visual exercise that proves that the interior sum of a triangle's angles is 180°.

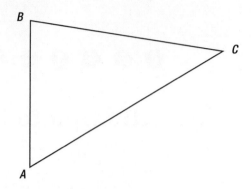

In this figure, $m\angle A + m\angle B + m\angle C = 180°$.

If you were to tear off the three corners of this triangle and line their untorn edges up next to one another, you would end up with a straight line, as shown in the following figure.

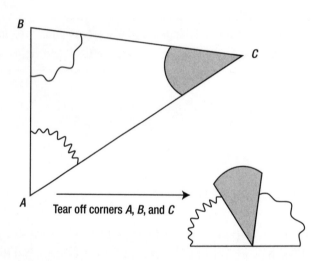

Tear off corners *A*, *B*, and *C*

Now, try this exercise yourself: Using a pair of scissors, cut off the corner of a piece of paper to make a large triangle. Next, carefully tear off the three corners. Finally, line up the untorn edges of each angle to see the straight angle that is formed!

TIP: The sum of the interior angles of a triangle is 180°.

The 180° sum of interior angles can be used to determine the measure of a missing angle from a triangle. Use this figure to understand the following two examples.

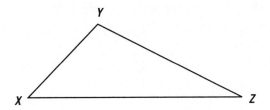

Example 1

If $m\angle X = 45°$ and $m\angle Y = 100°$, then the $m\angle Z$ can be determined by setting up $\angle X + \angle Y + \angle Z = 180°$. From this equation we see that $45° + 100° + \angle Z = 180°$. Therefore $\angle Z$ must equal $35°$.

Example 2

If $m\angle X = j$, $m\angle Y = 3j$, and $m\angle Z = 2j$, then the following equation can be set up: $\angle X + \angle Y + \angle Z = j + 3j + 2j = 6j = 180°$. From this equation we see that $j = 30°$, which means that $m\angle X = 30°$, $m\angle Y = 90°$, and $m\angle Z = 60°$.

PRACTICE 1

Use the following figure and the given angle information to answer questions 1–5.

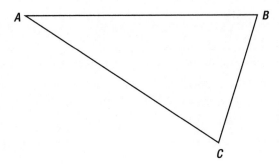

1. Find $m\angle A$ if $m\angle B = 72.3°$ and $m\angle C = 74.5°$.

2. Find the value of y if $m\angle A = (y + 9)°$, $m\angle B = (3y + 17)°$, and $m\angle C = (4y - 6)°$.

3. Using the information provided in question **2**, what is the measure of angles *A*, *B*, and *C*?

4. Find the measure of all the angles, given the following: The measure of angle *A* is five less than the measure of angle *B*. The measure of angle *C* is 15 less than twice the measure of angle *B*. (*Note:* Figure not drawn to scale.)

5. Assume that angle *C* is a right angle and angle *A* is one-half the measure of angle *B*. Find the measure of all the angles.

Naming and Labeling Triangles

Triangles are named using their three vertices, or angles. "Triangle *ABC*" can be written as △*ABC*. It is standard practice to name the angles in triangles with capital letters. The sides opposite each angle are named with the accompanying lowercase letter. For example, in the following figure, notice how side *l* is opposite ∠*L*, side *m* is opposite ∠*M*, and side *n* is opposite ∠*N*.

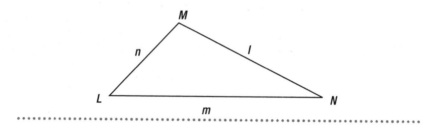

TIP: In triangles, the angles are named with capital letters and each opposite side is named with the corresponding lowercase letter.

Relationships between Sides and Angles in Triangles

There is a special relationship between the relative length of a side and the relative measurement of its opposite angle. (The word "relative" used in this way, means how it relates to the other parts of the triangle.) The longest side is always opposite the largest angle, the shortest side is always opposite the smallest angle, and the middle-length side is always opposite the middle-sized angle.

For example, if the previous figure were drawn to scale (with side m the longest, side n the shortest, and side l in between m and n), since side m is the longest side, we could conclude that $\angle M$ is the largest angle. Similarly, since side n is the shortest side, we could conclude that $\angle N$ is the smallest angle.

....

TIP: The largest angle in a triangle is always opposite its longest side. The smallest angle is always opposite the shortest side. The middle-sized angle is always opposite the middle-sized side.

....

Two-Side Sum Rule for Triangles

The sum of any two sides of a triangle must be greater than the length of the remaining side. For example, if two sides of a triangle were 7 and 10, the third side must be less than 17. If it were longer than 17, the other two sides would not be able to connect to the side's endpoints.

....

TIP: The sum of any two sides of a triangle must be greater than the length of the remaining third side.

....

PRACTICE 2

ΔNPR in the following figure is drawn to scale and should be used to answer questions **1–6**.

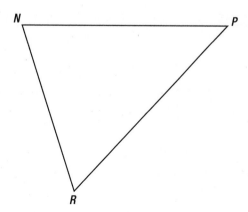

1. What would be another way to name side *NP*?

2. What would be another way to name side *PR*?

3. Given that $\triangle NPR$ is drawn to scale, which angle is the smallest and why?

4. If side $r = 17$ mm and side $p = 12$ mm, what is the largest even-numbered side length that could be possible for side *n*?

5. Given that $k > 0$, side $r = 6.5k + 0.5$, and side $n = 6.7k + 0.8$, which angle of $\triangle NPR$ must have the largest measure?

6. True or false: The side lengths for $\triangle NPR$ could be $n = 0.9$ cm, $r = 1.6$ cm, and $p = 0.7$ cm.

ACUTE, OBTUSE, AND EQUIANGULAR TRIANGLES

One way that triangles can be classified is by the measure of their angles. Read the following definitions and then look at the triangles in the following figure.

Acute triangles have three angles that each measure less than 90°.

Right triangles have one angle that measures 90°. (Why is it not possible for a triangle to have more than one angle that is 90°?) Since the sum of the interior angles of a triangle is 180°, the sum of the two acute angles in a right triangle is 90°.

Obtuse triangles have one angle that is greater than 90°. Because the sum of the interior angles of a triangle is 180°, only one angle in an obtuse triangle can be greater than 90°.

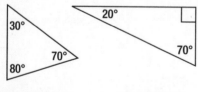
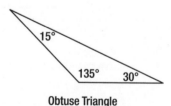

Acute Triangle   Right Triangle   Obtuse Triangle

PRACTICE 3

1. What is the largest integer-valued angle that an acute triangle can contain?

2. In $\triangle JOC$, $m\angle J = 52°$ and $m\angle O = 38°$. What kind of triangle is $\triangle JOC$?

3. In $\triangle FGH$, $FG < GH < HF$. Which angle is the largest?

4. Which of the following could *not* represent two angles from an acute triangle?
   a. 34° and 52°
   b. 18° and 62°
   c. 78° and 27°
   d. 61° and 29°

5. True or false: A right triangle can also be an acute triangle.

6. True or false: An obtuse triangle can have two obtuse angles.

7. True or false: An obtuse triangle can also be a right triangle.

8. In $\triangle TLC$, the measure of angle T is 48°. If the measure of $\angle L$ is three times the measure of $\angle C$, find the measure of angles L and C and classify what kind of triangle this is.

9. $\triangle MAT$ is a right triangle, with MA being perpendicular to AT. If $\angle T$ is 18 more than twice the value of $\angle M$, find the measures of $\angle T$ and $\angle M$.

ANSWERS

Practice 1

1. $m\angle A = 33.2°$, since $180° - 72.3° - 74.5° = 33.2°$.
2. $y = 20°$, since $(y + 9) + (3y + 17) + (4y - 6) = 180°$.
3. $m\angle A = 29°$, $m\angle B = 77°$, and $m\angle C = 74°$
4. $m\angle A = 45°$, $m\angle B = 50°$, and $m\angle C = 85°$
5. $m\angle A = 30°$, $m\angle B = 60°$, and $m\angle C = 90°$, since $90° + \frac{1}{2}B + B = 180°$.

Practice 2

1. side r (the lowercase of the opposite angle)
2. side n (the lowercase of the opposite angle)
3. $\angle P$ would be the smallest because it is opposite the shortest side.
4. Side n must be smaller than the sum of r and p, so the largest even-numbered side length it could be would be 28 mm.
5. Angle N must have the largest measure since it is opposite the largest side.
6. False. The sum of n and p is 1.6, which means that side r must be less than 1.6 cm.

Practice 3

1. $89°$
2. right triangle
3. $\angle G$
4. c. a, b, and d could not be angles from an acute triangle.
5. False
6. False
7. False
8. $m\angle L = 99°$, $m\angle C = 33°$, and $\triangle TLC$ is an obtuse triangle.
9. $\angle T = 66°$ and $\angle M = 24°$

a closer look at classifying triangles

Geometry existed before creation.

—PLATO

Now that you have learned the basics of triangles, it is time to learn about scalene, obtuse, and equiangular triangles. In this lesson you will also learn about a triangle's altitude, median, angle bisectors, and mid-segments.

CLASSIFYING TRIANGLES BY SIDE LENGTHS

In the previous lesson you learned how to classify triangles by their angle measurements. Triangles are also classified by their side lengths. Do you remember that there is a relationship between a triangle's side lengths and its angle measurements? When you are given information about a triangle's angle measurements, you can also identify information about their angles.

Scalene triangles have three side lengths that are all different. This means that their angles are also unique. Scalene triangles can either be **obtuse** or **acute**. See the next figure.

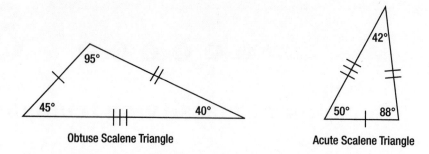

Obtuse Scalene Triangle   Acute Scalene Triangle

When a triangle has two angles that are equal in measure, it is called an **isosceles triangle**. The two congruent angles are referred to as the **base angles** and the side included between them is called the **base**. These congruent base angles are opposite a pair of congruent sides, which are called **legs**. The non-congruent angle is called the **vertex angle**. Isosceles triangles can either be obtuse or acute. See the next figure.

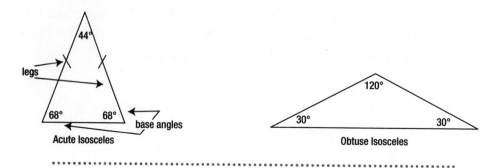

Acute Isosceles   Obtuse Isosceles

TIP: Isosceles triangles have one pair of congruent sides (called "legs") and one pair of congruent angles (called base angles).

Equilateral triangles** have three sides that are equal in length. Since they also have three equal angles that measure 60°, they are also referred to as **equiangular triangles**. Equiangular triangles are always acute. See the next figure.

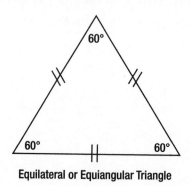

Equilateral or Equiangular Triangle

> **TIP:** The terms *equilateral* and *equiangular* both refer to triangles that have three equal angles of 60° and three congruent sides.

PRACTICE 1

1. In $\triangle ABC$, $m\angle A = 42°$ and $m\angle B = 96°$. First, identify the measure of $\angle C$ and then identify the most specific way to classify $\triangle ABC$.

2. In $\triangle EAT$, $m\angle E = 62°$ and $m\angle A = 88°$. First, identify the measure of $\angle T$ and then identify the most specific way to classify $\triangle EAT$.

3. In $\triangle MIC$, $\angle M$ is one-half the measure of $\angle I$, which is 90°. First, identify the measure of $\angle C$ and then determine the most specific way to classify $\triangle MIC$.

For questions **4–8**, identify whether the given conditions *sometimes*, *always*, or *never* result in an *acute* triangle.

4. An equiangular triangle with side length of 100 cm.

5. An isosceles triangle with a base angle measuring 15°.

6. A scalene triangle with one angle measuring 62°.

7. Isosceles $\triangle CAT$ where $\angle A$ is the vertex, and the vertex is three times larger than each base angle.

8. In $\triangle REM$, $RE = 10$ cm, $EM = 10$ cm, and $MR = 10$ cm.

CALCULATIONS WITH SPECIAL TRIANGLES

When you know that a triangle is isosceles, equilateral, or right, there are several ways you can identify missing information about its sides or angles.

Problem Solving with Isosceles Triangles

With an isosceles triangle, if you are given only one angle's measurement, it is possible to figure out the remaining two angles. You need only to know if the given angle is a base angle or a vertex angle.

Example 1 (vertex angle given)

Given that the vertex angle of an isosceles triangle is 120°, the remaining two base angles must equally share the remaining 60° of the triangle since 180° − 120° = 60°. In this case, the two base angles would each measure 30°.

Example 2 (base angle given)

Given that the base angle of an isosceles triangle is 75°, you know that the other base angle is also 75°. This would then leave 30° for the vertex of the triangle, since $180° − 2 \times (75°) = 30°$.

Algebraic Problem Solving with Isosceles Triangles

The same techniques can be used with isosceles triangles when you are given an expression in algebra to represent either the base or the vertex angle of an isosceles triangle.

Example 3 (algebraic expression given)

If given that the vertex angle of an isosceles triangle is 10 more than twice the measure of its base angle, the following could be identified: Each of the base angles could be represented with b and the vertex angle could be represented as $10 + 2b$. Then since the sum of the vertex angle plus two of the base angles is 180°, the following equation can be written: $(10 + 2b) + (b) + (b) = 180°$. Therefore, $10 + 4b = 180°$ and $b = 42.5°$. In this case, each base angle would measure 42.5° and the vertex angle would measure 95°.

PRACTICE 2

1. In an isosceles triangle, one of the base angles measures 37.2°. Find the measure of the vertex angle.

2. What is the measure of each base angle in an isosceles triangle whose vertex measures 117°?

3. In an isosceles triangle, the vertex angle measures $(5x - 10)°$ and each base angle measures $(y)°$. Write an algebraic expression to represent y in terms of x.

4. If the vertex angle of an isosceles triangle is four times as large as each of the base angles, find the measure of each angle of the triangle.

5. The base angle of an isosceles triangle is 30 less than twice the triangle's vertex angle. Find the measure of each of the angles and classify it as an acute or obtuse isosceles.

Problem Solving with Right Triangles

A helpful tidbit to remember about right triangles is that the two remaining acute angles will always be complementary. With this in mind, instead of summing up three angles to 180°, you can just sum the two acute angles to 90°.

Example 4 (right triangle)

In a right triangle, one of the acute angles is 14 less than three times the other acute angle. In order to find the acute angle measurements, let x represent one of the acute angles and let $3x - 14$ represent the larger acute angle. Then after solving the equation $(3x - 14) + (x) = 90°$, you will see that $x = 26°$. Therefore, the smaller acute angle will measure 26° and the larger one will be 64°.

Problem Solving with Equilateral Triangles

With equilateral or equiangular triangles, you can set the sides equal to one another to solve for variables when you are given algebraic expressions to represent the side lengths. The same is true for their angle measures. These can be set equal to one another.

Example 5 (equilateral triangle)

In an equilateral triangle, one of the sides measures $(4x - 20)$ centimeters and another side measures $(130 - 2x)$ centimeters. In order to find the side lengths, let the two sides equal each other in the equation: $4x - 20 = 130 - 2x$. Solving for x, you will get $x = 25$, which results in sides of 80 centimeters.

PRACTICE 3

1. A right triangle has one acute angle that measure 42.75°. Find the measurement of the remaining angle.

2. An equiangular triangle has one angle that measures $(5x - 12y)°$ and another angle that measures $(7x + 18y)°$. Write an expression that represents x in terms of y. (*Hint:* Write an equation that relates the two angles to each other, and then get x by itself.)

3. Use the following figure to solve for h, u, and f. (*Hint:* Notice what kind of triangle this is and begin by solving for h.)

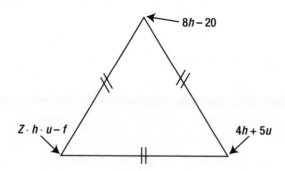

4. In an equiangular triangle, each of the angles equals $3p$. The side lengths are each $(2p - 7)$ inches. Find the measure of the angles and lengths of the sides.

ANSWERS

Practice 1

1. $m\angle C = 42°$ and $\triangle ABC$ is an obtuse isosceles triangle.
2. $m\angle T = 30°$ and $\triangle EAT$ is an acute scalene triangle.
3. $m\angle C = 45°$ and $\triangle MIC$ is a right isosceles triangle.
4. always (Regardless of its side length, an equiangular triangle will always be acute since its three angles will each measure 60 degrees.)
5. never
6. sometimes
7. never (The base angles would each measure 36° and the vertex angle would measure 118°.)
8. always (This is an equilateral triangle, which is always acute.)

Practice 2

1. 105.6°
2. 31.5°
3. $y = \frac{180 - (5x - 10)}{2}$
4. base angles = 30° and the vertex angle = 120°
5. base angles = 66° and the vertex angle = 48°, so this is an acute isosceles triangle.

Practice 3

1. 47.25° (90 − 42.75 = 47.25°)
2. Solve for x in the equation: $5x - 12y = 7x + 18y$

$$5x - 12y = 7x + 18y$$
$$5x - 7x = 18y + 12y$$
$$-2x = 30y$$
$$x = \frac{30y}{-2}$$
$$x = -15y$$

3. $h = 10$ (since $8h - 20 = 60°$), then $u = 4$ and $f = 20$
4. Each angle measures 60° so $p = 20$ and the side lengths are each 33 inches.

additional triangle properties

After finishing this lesson, you will have an understanding of how to identify and utilize a triangle's altitude, median, angle bisectors, and mid-segments. You will also understand the relationship between a triangle's external angle and remote interior angles.

FINDING A TRIANGLE'S ALTITUDE

The **altitude** of a triangle is one way of identifying a triangle's height in relation to a base side of the triangle. The altitude is the line drawn from any of the triangle's vertices to the opposite side. Because there are many ways that a line segment can be drawn from a vertex to the opposite side, there is a specific requirement that defines the altitude. A segment from a vertex to the opposite side is an altitude only if it forms a 90° angle with the opposite side. The length of the altitude is the triangle's height.

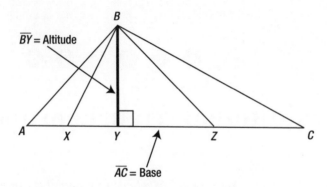

Looking at this figure, you can see that there are three line segments from $\angle B$ to side AC. Segments BX and BZ are not altitudes of $\triangle ABC$, since they are not perpendicular to AC. But line segment BY is the altitude of $\triangle ABC$, since it is perpendicular to side AC. The side intersected by the altitude is referred to as the **base**.

....

TIP: The *altitude* of a triangle is the line segment drawn from one vertex to the opposite side, intersecting the opposite side to form a right angle.

....

Nonstandard Altitudes

The altitude does not have to be drawn from the top vertex to the bottom side. It can be drawn in different directions as you can see in this figure.

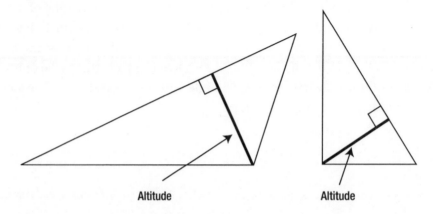

When you are finding the altitude of an obtuse triangle, sometimes the base must be extended so that the altitude can intersect the base in a perpendicular manner. You can see this in the next figure. Notice that in ΔJEN, side JN is extended out to point P so that the altitude EP can be drawn.

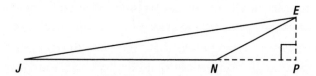

The last important concept to understand about altitudes is that triangles always have three altitudes—one from each vertex. In obtuse triangles the sides will need to be extended in order to draw the altitudes (as in the previous figure), but in acute triangles the three altitudes will intersect at a common point inside the triangle as in the following figure.

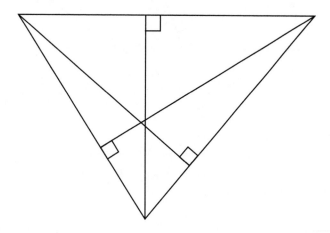

PRACTICE 1

Identify each statement below as *true* or *false*.

1. More than one altitude can be drawn from the vertex of a triangle.

2. Some triangles have only one altitude.

3. Sometimes the altitude extends outside of a triangle.

4. The *base* is the side of the triangle that the altitude intersects at a right angle.

5. In the first figure on altitudes the altitude can be either segment EP or segment JP.

MEDIANS, ANGLE BISECTORS, AND MID-SEGMENTS

The **median** of a triangle is similar to the altitude. It is drawn from the vertex to the opposite side. However, instead of intersecting the opposite side at a 90° angle, it intersects the opposite side at its midpoint and divides the side into two equal segments. Every triangle has three medians that intersect at a common point inside the triangle. Remember that identical hash marks are used to indicate congruent segments. Notice that in the following figure segments *XA* and *YA* are congruent, as well as segments *XC* and *WC*, and segments *WB* and *BY*. *Note:* Although the median divides the opposite side into congruent parts, it does not necessarily create two congruent angles.

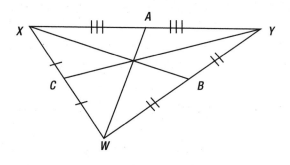

TIP: A *median* of a triangle is the segment drawn from one vertex to the midpoint of the opposite side.

You probably remember that the word *bisect* means *to cut something in half*. An **angle bisector** is the segment drawn from a vertex of a triangle to the opposite side so that the vertex is divided into two congruent angles. As with medians, every triangle has three angle bisectors that intersect at one common point inside the triangle. The standard way to show that two angles are congruent is to make a small arc in each angle and put the same number of hash marks through each arc. You can see these markings in the following figure. *Note:* Although an angle bisector divides the angle into two congruent angles, it does not necessarily divide the opposite side into two congruent segments.

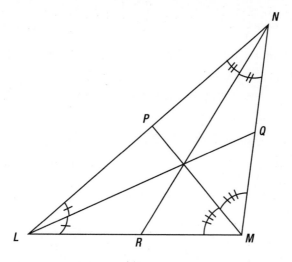

....

TIP: A triangle's *angle bisector* is a segment that divides any vertex angle into two congruent angles and extends to the opposite side.

....

Last, a triangle's **mid-segment** is a segment that connects the midpoints of any two sides. The mid-segment is always *parallel* to the side it is not connected to *and* it is also *half the length* of this non-connected side. Every triangle has three mid-segments, each one parallel to the third side and half its length, as shown in the following figure. In triangle *CAT*, *DG* is a mid-segment parallel to *CA*. *CA* is 22 units long and *DG* is 11 units long.

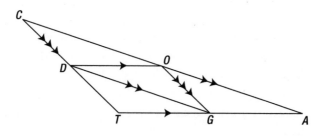

....

TIP: A triangle's *mid-segment* is a segment that connects the two midpoints of any two sides. It is parallel to the third side and half its length.

....

PRACTICE 2

1. If in the median figure $m\angle WXY$ is 50°, what is the $m\angle WXB$?

2. If in the angle bisector figure $m\angle LMN$ is 110°, what is the $m\angle LMP$?

3. If in the angle bisector figure the $m\angle NLQ$ is 24° and the $m\angle LMP$ is 47°, what is the $m\angle MNR$?

4. In the mid-segment figure, what line segment is parallel to side CT?

5. In the mid-segment figure, if side CA is 12.6 cm, the length of what other line segment can be determined? What is its length?

6. For what type of triangle do the three medians intersect at a common point *inside* the triangle?

7. Given an acute triangle RGB, if RX is the altitude, what can be stated about RX?

8. True or false: A triangle's angle bisectors will always cut the angles and opposite sides in half.

EXTERIOR AND INTERIOR ANGLES IN TRIANGLES

So far our exploration of triangles has been limited to interior angles. Triangles also have angles that occur outside of the triangle, when one of the sides is extended. In the following figure, $\angle 1$ is defined as the **exterior angle** of $\triangle ABC$. Since $\angle 2$ is next to $\angle 1$, we call this the **adjacent interior angle**. $\angle 3$ and $\angle 4$ are not adjacent to $\angle 1$, so they are considered the **remote interior angles** to $\angle 1$.

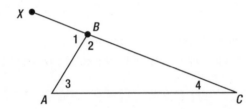

The exterior angle of a triangle has a special relationship to its remote interior angles:

You can see that $m\angle2 + m\angle1 = 180°$, since together they form a straight angle. You also remember that $m\angle2 + m\angle3 + m\angle4 = 180°$.

Using substitution with the two previous equations, use substitution to conclude that $m\angle1 = m\angle3 + m\angle4$. This relationship states that the measure of a triangle's exterior angle is equal to the sum of its two remote interior angles.

....

TIP: The measure of a triangle's *exterior angle* is always equal to the sum of the measures of its two remote interior angles.

....

PRACTICE 3

Use the preceding figure to answer questions **1–3**. Assume that the figure is *not* drawn to scale.

1. If $m\angle1 = 88°$ and $m\angle3 = 49°$, what is the $m\angle4$?

2. If $m\angle1 = g°$ and $m\angle4 = h°$, what algebraic expression would represent $m\angle4$? (*Hint:* How would you identify the $m\angle4$? Leave your answer in terms of g and h.)

3. If $m\angle1 = (3x - 12)°$, $m\angle4 = (2x - 4)°$, and $m\angle3 = (0.5x + 8.5)°$, find the measure of all the angles in $\triangle ABC$.

4. True or false: A triangle's exterior angle will always be obtuse.

5. An isosceles triangle has a base angle R that measures 37°. An exterior angle is drawn next to $\angle R$. What is the measure of this exterior angle?

ANSWERS

Practice 1

1. False
2. False
3. True
4. True
5. True

Practice 2

1. It cannot be determined. Only the sides are bisected by a triangle's medians.
2. $m\angle LMP = 55°$
3. 19°: In this figure, since $m\angle NLQ$ is 24°, then $m\angle NLM$ is 48°. Since $m\angle LMP$ is 47°, then $m\angle LMN = 94°$. Therefore, the $m\angle MNL$ will be $(180 - 48 - 94) = 38°$. $m\angle MNR$ will be half of this, so $m\angle MNR = 19°$.
4. Line segment OG
5. It can be determined that segments DG, CO, and OA would all measure 6.3 cm.
6. acute triangle
7. Segment RX intersects side GB at a right angle and is the height of the triangle.
8. False. Angle bisectors will always cut a triangle's angles in half, but they will not always divide the opposite side into two equal parts.

Practice 3

1. $m\angle 4 = 39°$
2. $m\angle 4 = (g - h)$. ($m\angle 1 = m\angle 3 + m\angle 4$, which means $g = h + m\angle 4$)
3. $x = 33°$, so $m\angle 1 = 87°$, $m\angle 3 = 62°$, $m\angle 4 = 25°$, and $m\angle 2 = 93°$
4. False. A remote angle will be acute when it is adjacent to an obtuse angle, and it will be right when it is adjacent to a 90-degree angle.
5. The exterior angle will measure 143°. (Both base angles will measure 37°, which leaves 106° for the vertex angle. The exterior angle will equal the sum of the two remote interior angles, which is 143°.)

perimeter and area of triangles

Inspiration is needed in geometry, just as much as in poetry.
—ALEKSANDR PUSHKIN

You are now ready to learn how to find the perimeter and area of right, acute, and obtuse triangles.

PERIMETER OF TRIANGLES

The concept of perimeter is something that is used in the real world every day. The word *perimeter* means *outer edge*. The perimeter of a soccer field is always marked with a painted out-of-bounds line. The perimeter of your school's property might be lined with a fence. The perimeter of any polygon is calculated by measuring the distance around an object. The **perimeter** of a triangle is the sum of the lengths of its three sides.

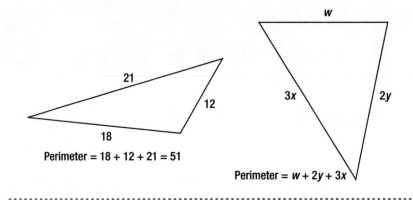

Perimeter = 18 + 12 + 21 = 51

Perimeter = $w + 2y + 3x$

TIP: The *perimeter* of a triangle is the sum of its three sides.

PRACTICE 1

1. The three sides of a triangle are consecutive odd integers. If the smallest side of the triangle is 11 cm, what is the perimeter of the triangle?

2. A corner garden plot has sides that measure 9 feet, by 12 feet, by 15 feet. If Peter wants to buy some short fencing to keep the rabbits out, how many feet of fencing does he need to purchase?

3. What is the perimeter of an equilateral triangle that has a side length of 7.5?

4. Ty's Tiles is having a one-day sale on their Mediterranean Tile, which is sold by the vertical foot. Veronica is in the store and wants to buy tile for a corner table that she just made. The table is an isosceles triangle and she remembers cutting the wood pieces for the congruent sides of the triangle to 4 feet, 8 inches. She also remembers that the non-congruent side of the triangle was 6 feet, 4 inches. What is the amount of tile that Veronica needs to purchase, in feet and inches, if she wants to line the perimeter of the table with Mediterranean Tile?

5. Find the perimeter in terms of x of this figure.

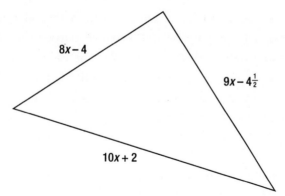

6. Using the same figure, find the perimeter of the triangle if $x = 2.5$ mm.

7. What is the side length of an equilateral triangle whose perimeter is 19.5 feet?

AREA

You just learned that the perimeter of a polygon is the distance around its *outside* edge. Another common measure of polygons is its **area**. The area of a polygon is the amount of space *inside* it. Area is measured in terms of the number of squares that are needed to fully cover a polygon. These squares can be *units*, or can be square inches, feet, or even miles. There is not one square unit that is *more* correct than the others in all cases—which one to use depends on the type of shape being discussed. For example, a large plot of land might be measured in square miles, but it would be better to measure a garden plot in square feet or square yards. The correct way area is written is with a squared exponent following the unit of measure, such as 25 in.^{2}. The squared exponent is referring to the inches and the 2 is NOT applied to the 25; 25 in.^{2} is said *25 square inches*. An area of 25 in.^{2} means that 25 one-by-one-inch squares are needed to fill a certain space.

....

TIP: *Area* is the measurement of the amount of space contained inside a figure. It is measured in square units and written with an exponent of 2 following the units.

....

Finding the Area of a Triangle

You probably remember from the previous lesson that the **altitude** of a triangle is considered to be that triangle's **height**. You also learned that the side that the altitude was perpendicularly drawn to is called the **base**. (Remember that any side can be the base as long as a perpendicular altitude is drawn to it.) The formula for calculating the area of a triangle is $A = \frac{1}{2}bh$, where A = the area, b = the length of the base, and h = the height.

....

TIP: The area of a triangle is given by $A = \frac{1}{2}(base)(height)$

....

Note that in the next figure, by using different sides as the *base*, there are three different ways that the area can be correctly calculated:

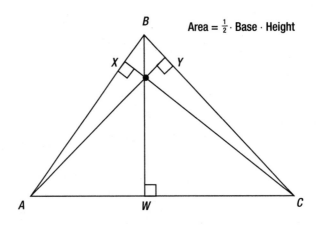

Area = $\frac{1}{2}$ · Base · Height

$A = \frac{1}{2}\,\overline{AC} \cdot \overline{BW}$ (using $\overline{AC}$ as base)
or
$A = \frac{1}{2}\,\overline{AB} \cdot \overline{CX}$ (using $\overline{AB}$ as base)
or
$A = \frac{1}{2}\,\overline{BC} \cdot \overline{AY}$ (using $\overline{BC}$ as base)

Finding the Area of an Obtuse Triangle

When you are working with obtuse triangles, the same formula for area is used, but take special care to choose the correct pairing of altitude and base. You already know that in obtuse triangles, sometimes the base must be extended so that a perpendicular altitude can be drawn from the opposite vertex. It is important to remember to use the length of the external altitude as your height, and the *non*-extended base as your base. For example, in the following figure, *JK* is the *non*-extended base that you will use to calculate the area (NOT *JE*).

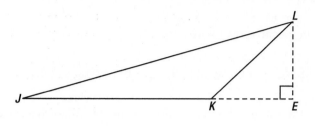

$$A = \tfrac{1}{2}\overline{JK} \cdot \overline{LE}$$

PRACTICE 2

Use this figure to answer questions 1–7.

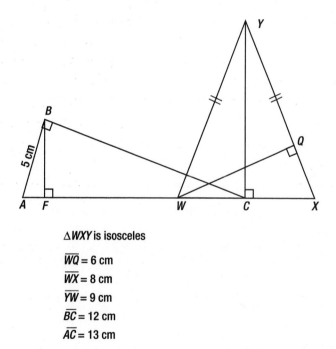

△*WXY* is isosceles

$\overline{WQ}$ = 6 cm

$\overline{WX}$ = 8 cm

$\overline{YW}$ = 9 cm

$\overline{BC}$ = 12 cm

$\overline{AC}$ = 13 cm

1. What is the area of △*WXY*? (*Hint:* Notice that △*WXY* is isosceles.)

2. Use your answer from question **1** to find the length of altitude *YC*.

3. What is the area of △*ABC*?

4. Use your answer from question **3** to find the altitude *BF*.

5. What is the perimeter of $\triangle WXY$?

6. What is the perimeter of $\triangle ABC$?

7. If segment $WC = 4$ cm, what is the length of edge AX?

State whether the statements in questions **8–10** are sometimes, always, or never true.

8. If the perimeter of $\triangle ANT$ is greater than the perimeter of $\triangle DOG$, then $\triangle ANT$ must have a greater area.

9. Perimeter is measured in square units, with examples like in.^{2}, ft.^{2}, or km^2.

10. Every triangle has three altitudes and three bases.

ANSWERS

Practice 1

1. 39 cm $(11 + 13 + 15 = 39)$
2. 36 feet
3. 22.5
4. 15 feet, 8 inches. (4 ft. 8 in. + 4 ft. 8 in. + 6 ft. 4 in. = 14 ft. 20 in. = 14 ft. + (1 ft. 8 in.) = 15 ft. 8 in.)
5. $27x - 6\frac{1}{2}$
6. 61 mm
7. 6.5 ft. $(\frac{19.5}{3} = 6.5)$

Practice 2

1. 27 cm^2
2. $\frac{27}{4}$ or 6.75 cm. (Solve for the height in this equation for area: $27 = \frac{1}{2}(8)(\text{height})$)
3. 30 cm^2
4. $\frac{60}{13}$ cm. (Solve for the height in this equation for area: $30 = \frac{1}{2}(13)(\text{height})$)

5. 26 cm

6. 30 cm (5 + 12 + 13 = 30)

7. Segment AX = 17 cm (13 + 8 − 4 = 17)

8. Sometimes. (Think about this: A long skinny triangle will have a large perimeter, but not a large area!)

9. Never. (Area is measured in square units, but perimeter is NOT.)

10. Always. (Although in isosceles triangles the sides will need to be extended to draw the altitude, they can still be drawn!)

the Pythagorean theorem

I never got a pass mark in math . . . Just imagine—
mathematicians now use my prints to illustrate their books.
—M. C. Escher

In this lesson you will learn about the widely used Pythagorean theorem and its application to right triangles.

A CLOSER LOOK AT RIGHT TRIANGLES

So far you have learned that a right triangle has one right angle and two acute angles that sum to 90°. Before discussing the Pythagorean theorem, you must become familiar with the standard way to label the sides and angles of right triangles. In a right triangle, the two sides that form the right angle are referred to as the **legs**. The longest side is always opposite the right angle and is called the **hypotenuse**. It is standard practice to label the legs of a right triangle a and b, while the hypotenuse is labeled c. As discussed in previous lessons, the angles across from the sides are named with the capital letters of their opposite sides. Note in the next figure that the right angle is labeled C and the two acute angles are labeled A and B.

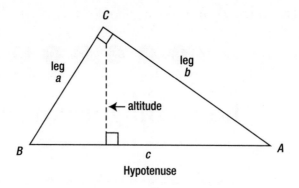

TIP: In a right triangle, the shorter sides are called the *legs* and are labeled *a* and *b*. The longest side opposite the right angle is called the *hypotenuse* and is labeled *c*.

THE PYTHAGOREAN THEOREM

The Pythagorean theorem is used to find the missing third side length in a right triangle when you are given any two sides. It is common to be given the lengths of the two legs, and to then use the Pythagorean theorem to solve for the hypotenuse. However, it is just as easy to use the Pythagorean theorem to solve for the length of one of the legs, when you're given the length of the other leg and the hypotenuse. The Pythagorean theorem is:

$$(\text{Leg } \#1)^2 + (\text{Leg } \#2)^2 = (\text{Hypotenuse})^2$$

Since the two legs are commonly labeled *a* and *b*, and the hypotenuse is commonly labeled *c*, the Pythagorean theorem is most commonly written as.

$$a^2 + b^2 = c^2$$

TIP: The Pythagorean theorem applying to the sides of right triangles is $a^2 + b^2 = c^2$.

Solving for the Hypotenuse Using the Pythagorean Theorem

Using the following figure, we will demonstrate how to solve for the missing hypotenuse using the Pythagorean theorem. (*Note:* It does not matter which leg you chose to label *a* and which leg you chose to label *b*.)

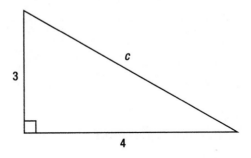

$a^2 + b^2 = c^2$

$3^2 + 4^2 = c^2$

$9 + 16 = c^2$

$25 = c^2$

$\sqrt{25} = \sqrt{c^2}$

$5 = c$, so the hypotenuse *c* in this example equals 5.

Solving for a Leg Using the Pythagorean Theorem

Using the following figure, we will demonstrate how to solve for a missing leg when the hypotenuse is one of the given pieces of information.

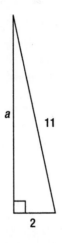

$$a^2 + b^2 = c^2$$

$$a^2 + 2^2 = 11^2$$

$$a^2 + 4 = 121$$

$$a^2 = 117$$

$$\sqrt{a^2} = \sqrt{117}$$

$a = \sqrt{117} \approx 10.8$. So the missing leg, a in this example, approximates to 10.8.

Be Careful of This Common Mistake!

The missing piece of information must be correctly identified and then shown in your equation when you are working with right triangles. A common mistake with the Pythagorean theorem is when students assume that whatever side is missing should be represented with c. You must be careful to notice whether you are solving for a missing leg (a or b) or a missing hypotenuse (c).

PRACTICE 1

Use this figure to answer questions **1–4**.

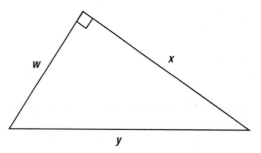

Note: Figure Not Drawn to Scale

1. Solve for y if $x = 8$ and $w = 6$.

2. Solve for x if $y = 13$ and $w = 5$.

3. Solve for w if $y = 10$ and $w = 5$.

4. Solve for y if $x = 4k$ and $w = 2k$.

5. Victoria has a shed that is 15 feet long and 10 feet wide. She wants to store the longest rolled-up carpet possible in the shed, diagonally on the floor. What is the largest rolled carpet that will fit in Victoria's shed?

6. Dorothy is standing directly 300 meters under a plane. She sees another plane flying straight behind the first. It is diagonally 500 meters away from her. How far apart are the planes from each other?

7. Eva and Carr meet at a corner. Eva turns 90° left and walks 5 paces; Carr continues straight and walks 6 paces. If a line segment connected them, how long would it measure?

OTHER APPLICATIONS OF THE PYTHAGOREAN THEOREM

Proving Right Triangles

Since the relationship between the legs and the hypotenuse in right triangles is $a^2 + b^2 = c^2$, we can work backwards to see if a given triangle is right or not. In the following figure, we can test to see if the lengths of the sides of the given triangle identify the Pythagorean theorem. If they result in a balanced equation, where the left side is equal to the right side, then $\triangle PIG$ is a right triangle. If the equation is not balanced, then you know that $\triangle PIG$ is not right.

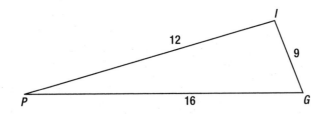

$a^2 + b^2 = c^2$

$9^2 + 12^2 = 16^2$ (Use the longest side as the hypotenuse.)

$81 + 144 = 256$

$225 \neq 256$

Since the result is a false statement, you can conclude that $\triangle PIG$ is not right.

Identifying Acute and Obtuse Triangles

Similarly, when you are given the length of all three sides of a triangle, you can use the Pythagorean theorem to identify whether it is an acute or an obtuse triangle. If the hypotenuse squared is larger than the sum of the squared legs, then the triangle is obtuse. See the next figure for an example that illustrates this.

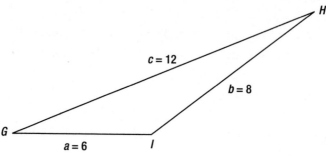

$$a^2 + b^2 = c^2$$

$$6^2 + 8^2 = 12^2$$

$$36 + 64 = 144$$

$$100 < 144$$

Therefore, $\triangle GHI$ is obtuse.

Conversely, if the hypotenuse squared is smaller than the sum of the squared legs, then the triangle is acute. See the next figure for an example that illustrates this.

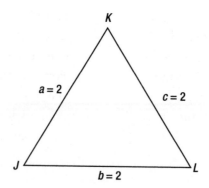

$$a^2 + b^2 = c^2$$

$$2^2 + 2^2 = 2^2$$

$$4 + 4 = 4$$

$$8 > 4$$

Therefore, $\triangle JKL$ is acute.

PRACTICE 2

For questions **1–5**, given the following side lengths, identify whether the triangle is right, acute, or obtuse:

1. 12, 16, 20

2. 8, 9, 12

3. 4.6, 7.8, 11.5

4. 11, 11, 16

5. $2\sqrt{3}$, 4, $2\sqrt{7}$

THE 3-4-5 RIGHT TRIANGLE

One helpful ratio to memorize with right triangles is the 3-4-5 triangle. A quick calculation using the Pythagorean theorem shows that a 3-4-5 triangle is right, since $9 + 16 = 25$. Any triangle that has sides that can be reduced to a ratio of 3:4:5 will be a right triangle. For example, if you are given a triangle with side lengths of 30, 40, and 50, you might recognize that when you divide all the sides by 10, the ratio of the sides is 3:4:5. Seeing this, you can conclude that a triangle with sides 30, 40, and 50 is a right triangle.

The 3-4-5 property can also be used as a shortcut to the Pythagorean theorem. If you see that two of them can be reduced to fit the 3:4 ratio, then you can use this shortcut to determine the missing hypotenuse without going through the calculations of the Pythagorean theorem. For example, if you know that a right triangle has two legs that measure 6 and 8, you might notice that since $3 \times 2 = 6$ and $4 \times 2 = 8$, a multiple of 2 can be used to identify the hypotenuse by computing $5 \times 2 = 10$. Similarly, if you are given a leg and a hypotenuse that can be reduced to a 3:5 or 4:5 ratio, the same shortcut can be used. Consider a right triangle with a leg of 33 and a hypotenuse of 55. Since $3 \times 11 = 33$ and $5 \times 11 = 55$, a multiple of 11 can be used to determine that the second leg would be $4 \times 11 = 44$.

PRACTICE 3

Given the two sides of right triangles, use the 3-4-5 ratio to find the missing side.

1. leg 1 = 12, leg 2 = 16; find the length of the hypotenuse.

2. leg 1 = 18, leg 2 = 24; find the length of the hypotenuse.

3. leg 1 = 21, hypotenuse = 35; find the length of leg 2.

4. leg 1 = 1.5, hypotenuse = 2.5; find the length of leg 2.

5. leg 1 = $3xy$, leg 2 = $4xy$; find the length of the hypotenuse.

ANSWERS

Practice 1

1. $y = 10$
2. $x = 12$
3. $w = 5\sqrt{3}\left(\sqrt{75} = \sqrt{25 \times 3} = \sqrt{25} \times \sqrt{3} = 5\sqrt{3}\right)$
4. $y = 2k\sqrt{5}$ $\left(16x^2 + 4x^2 = 20x^2 = \sqrt{4 \times 5 \times k^2} = 2k\sqrt{5}\right)$
5. 18 feet ($\sqrt{325} \approx 18.03$)
6. 400 meters
7. $\sqrt{61} \approx 7.8$

Practice 2

1. right
2. acute
3. obtuse
4. obtuse
5. right

Practice 3

1. 20
2. 30
3. 28
4. 2
5. $5xy$

special right-triangle relationships

Geometry is the science of correct reasoning on incorrect figures.
—GEORGE PÓLYA

In this lesson you will apply what you learned about the Pythagorean theorem to the special side–angle relationships in 30°-60°-90° triangles and 45°-45°-90° triangles.

45-45-90 RIGHT TRIANGLES

There are two common right triangles that have special angle–side relationships. The first noteworthy triangle is the **isosceles right triangle**, which is also referred to as the **45-45-90 right triangle**. As you remember, isosceles triangles have one vertex angle and two congruent base angles. If the vertex angle is 90°, the two base angles must sum to 90°, so they each measure 45°. That is why this special isosceles triangle is called the 45-45-90 right triangle. It is pictured in the next figure.

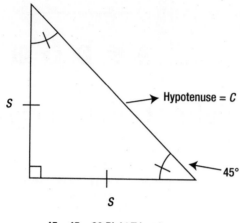

45 – 45 – 90 Right Triangle

The Pythagorean Theorem and the 45-45-90 Right Triangle

Since the legs of an isosceles triangle are congruent, they can each be labeled s instead of using a and b. The hypotenuse is labeled c. Then, applying the Pythagorean theorem, we can solve for c. This will reveal the special relationship between the legs and the hypotenuse in 45-45-90 right triangles:

$$s^2 + s^2 = c^2$$

$$2s^2 = c^2$$

$$\sqrt{2s^2} = \sqrt{c^2}$$

$s\sqrt{2} = c$, so the hypotenuse c will always be s times $\sqrt{2}$ (which is the length of one leg times $\sqrt{2}$).

....

TIP: In 45-45-90 right triangles, the length of the hypotenuse is equal to $\sqrt{2}$ times the length of one of the legs. This is written $s\sqrt{2}$.

....

Working the other direction, it is possible to see that if you were given the length of the hypotenuse, you could divide it by $\sqrt{2}$ to get the length of one of the legs. However, because it is not acceptable to keep a radical sign in the denominator of a fraction, another way to write this leg is $= \left(\dfrac{\text{hypotenuse}}{2} \right)(\sqrt{2})$.

TIP: In 45-45-90 right triangles, the length of one of the congruent legs is equal to the hypotenuse times $(\frac{\sqrt{2}}{2})$. This is written $\frac{c}{2} \times \sqrt{2}$.

PRACTICE 1

Use the following figure to solve questions **1–6**. (*Notice:* The sides are not labeled with the standard letters. This does not change how the rules for 45-45-90 right triangles apply.)

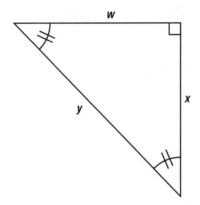

1. If $w = 12$, find the length of y.

2. If $y = 32$, find the length of x.

3. If $x = 5\sqrt{2}$, find the length of y.

4. If $y = 14\sqrt{6}$, find the length of x.

5. If $x = 7.5$, find the perimeter of the triangle to the nearest tenth.

6. If $y = (8\sqrt{2})g$, find the perimeter of the triangle in terms of g.

7. The perimeter of a 45-45-90 isosceles right triangle is 41 feet. Find the length of the triangle's legs and hypotenuse. (*Hint:* Assign variables to the legs and hypotenuse that reflect their relationship in 45-45-90 triangles, and then use those variable expressions in the perimeter formula.)

8. Izzy is making a gardening table from recycled wood to fit in the back corner of his yard. He wants the table to be an isosceles right triangle with the base edge facing out toward the garden. The longest piece of wood he has for the front edge of the table (the base of the triangle) is 5 feet, 4 inches. How many inches long will each of the congruent sides of the table be?

9. Angle *T* of right triangle *TUV* measures 45°. If base *TU* measures 10 units, what is the length of the hypotenuse of triangle *TUV*?

10. Shelly is 40 miles directly north of Bend, Oregon. Adam is 40 miles directly east of Bend, Oregon. What is the shortest distance, diagonally, between Shelly and Adam? (Round to the nearest tenth of a mile.)

30-60-90 RIGHT TRIANGLES

The second common right triangle with special angle–side relationships is the **30-60-90 right triangle**. The 30-60-90 right triangle has a right angle that measures 90°, and two acute angles that measure 30° and 60°. As you remember, the smallest side of a triangle is directly opposite the smallest angle. Therefore, we are going to call the length of the shortest leg opposite the 30° angle, *n*. The side opposite the 60° angle in 30-60-90 right triangles always has a length of $\sqrt{3}$ *times the shortest side* or $n\sqrt{3}$. The hypotenuse in 30-60-90 right triangles is always *twice the length of the shorter side*, or 2*n*. Notice how the next figure is labeled.

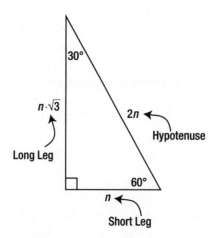

TIP: In 30-60-90 right triangles, if the length of the smallest leg is n, then the length of the longer leg is $n\sqrt{3}$ and the length of the hypotenuse is $2n$.

The Pythagorean Theorem and the 30-60-90 Right Triangle

The Pythagorean theorem can be used to confirm the special relationship between the sides of the 30-60-90 right triangle.

$a^2 + b^2 = c^2$

$(n)^2 + (n\sqrt{3})^2 = (2n)^2$

$n^2 + 3n^2 = 4n^2$

$4n^2 = 4n^2$ and since this is a true statement, the relationship between sides is confirmed to be correct.

Working Backwards with 30-60-90 Triangles

If you are given the length of the hypotenuse in a 30-60-90 right triangle, then you can calculate the length of the shorter side by dividing it by 2. Similarly, if you're given the length of the longer leg, you can calculate the length of the shorter side by dividing it by $\sqrt{3}$. If you are not given the length of the shorter side, it is easiest first to solve for the shorter side and then to move forward from there.

Example

The length of the longer leg of a 30-60-90 right triangle is 10. Find the length of the shorter leg and of the hypotenuse.

We know that the (shorter leg)($\sqrt{3}$) = (long leg).

So the (shorter leg)($\sqrt{3}$) = 10.

Therefore, the shorter leg = $\frac{10}{\sqrt{3}} \approx 5.8$.

Since the hypotenuse = (shorter leg)(2), the hypotenuse = 5.8(2) = 11.6.

PRACTICE 2

Use the next figure to answer questions 1–5.

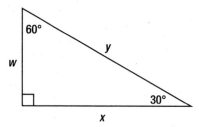

1. If $w = 7.5$, find the length of y.

2. If $y = 36$, find the length of w.

3. If $x = 6\sqrt{3}$, find the length of y.

4. If $y = 12\sqrt{10}$, find the length of x.

5. If $w = 22$, find the perimeter of the triangle to the nearest tenth.

6. $\triangle PAT$ is a 30-60-90 right triangle with the right angle at point A. If PA is the shorter leg, and $AT = 64$, find the length of the hypotenuse to the nearest tenth.

7. $\triangle SKY$ is a 30-60-90 right triangle with the right angle at point Y. If $SK = 38$ inches, find the perimeter to the nearest tenth.

Use the next figure to answer questions 8–10.

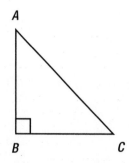

8. If the length of side AB is 9 cm, what is the length of side BC?

9. If the hypotenuse of triangle ABC is $6x + 2$ units long, what is the length of side AB?

10. If the sum of sides BC and AC is 12 units, what is the length of side AB?

ANSWERS

Practice 1

1. $y = 12\sqrt{2} \approx 16.8$ units
2. $x = 16\sqrt{2} \approx 22.6$ units
3. $y = 10$ units ($5\sqrt{2} \times \sqrt{2} = 5 \times 2 = 10$)
4. $y = 14\sqrt{6}$, and since this is the hypotenuse, multiply it by $\frac{\sqrt{2}}{2}$:

$$\frac{\sqrt{2}}{2} \times \frac{14\sqrt{6}}{1}$$

$$\frac{14\sqrt{12}}{2}$$

$$7\sqrt{12} = 7\sqrt{4}\sqrt{3} = 14\sqrt{3} \approx 24.2$$

5. $P \approx 25.6$ units ($P = 7.5 + 7.5 + 7.5\sqrt{2} \approx 25.6$)
6. $P \approx 27.3g$ units ($P = y + w + x = (8\sqrt{2})g + \frac{(8\sqrt{2})g(\sqrt{2})}{2} + \frac{(8\sqrt{2})g(\sqrt{2})}{2} = (8\sqrt{2})g + (16g) \approx 27.3g$)
7. Since it is a 45-45-90 isosceles right triangle, you can let each of the legs measure x units and, therefore, the hypotenuse will then measure $x\sqrt{2}$. The equation

$$P = x + x + x\sqrt{2} = 41$$
$$1x + 1x + 1.41x = 41$$
$$3.41x = 41, \text{ so } x \approx 12$$

Therefore, each of the legs is 12 and the hypotenuse is $12\sqrt{2}$.
8. The base is 5 feet 4 inches, which is $5(12) + 4 = 64$ inches. In order to find the length of each of the congruent sides, multiply 64 by ($\frac{\sqrt{2}}{2}$): $64(\frac{\sqrt{2}}{2}) = 32(\sqrt{2}) \approx 45.25$ inches
9. The legs are each 8 units and the hypotenuse is 11.3 units. (Solve the equation $x + x + x\sqrt{2} = 27.1$)
10. 56.6 miles

Practice 2

1. $y = 15$ units
2. $w = 18$ units
3. $y = 12$ units ($\frac{6\sqrt{3}}{\sqrt{3}} = 6$, which is the shorter side, then $6 \times 2 = 12$)
4. $x = 32.9$ units ($\frac{12\sqrt{10}}{2} = 6\sqrt{10}$, which is the shorter leg. Then the longer leg is $6\sqrt{10} \times \sqrt{3} = 6\sqrt{30} \approx 32.9$ units)
5. $P \approx 104.1$ units ($P = 22 + 44 + 22\sqrt{3} \approx 25.6$)
6. Hypotenuse ≈ 73.9 units ($\frac{64}{\sqrt{3}} \approx 36.95$, which is the shorter side; then $36.95 \times 2 = 73.9$)
7. $P \approx 89.9$ inches ($P = 38 + 19 + 19\sqrt{3} \approx 89.9$)
8. $3\sqrt{3} \approx 5.2$ ($\frac{9}{\sqrt{3}} = \frac{9\sqrt{3}}{3} = 3\sqrt{3}$)
9. $(3x + 1)\sqrt{3}$ units. ($\frac{6x+2}{2} = (3x + 1)$ which is the shorter leg. Then multiplying that by) $\sqrt{3}$ equals the length of the longer leg)
10. $4\sqrt{3}$ units. ($12 = $ shorter leg (n) + hypotenuse ($2n$). So $12 = 3n$ and the shorter leg $= 4$.)

triangle congruence

The essence of mathematics is not to make simple things complicated, but to make complicated things simple.

—S. GUDDER

In this lesson you will learn what congruence is and how to identify if two triangles are congruent.

WHAT IS CONGRUENCE?

When two shapes are congruent, it means that they have the same appearance, size, and relationships between their different parts. For example, if a tire on a small passenger car goes flat, a tire that is congruent to the damaged tire is needed to replace it. An oversized tire from a tractor could not be used as a replacement, nor could a thin tire from a bicycle. The replacement tire would need to have the same thickness, height, and overall shape; it would need to be congruent to the original tire. In geometry there are very specific rules that define whether two shapes are congruent:

Line segments are congruent when they are the same length, regardless of how they are arranged on paper.

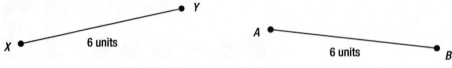

$\overline{XY} \cong \overline{AB}$: $\overline{XY}$ is congruent to $\overline{AB}$

Angles are congruent when they contain the same number of degrees. They do not need to be facing the same direction in order to be congruent.

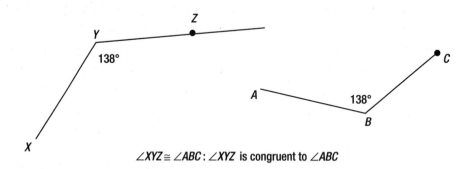

$\angle XYZ \cong \angle ABC$: $\angle XYZ$ is congruent to $\angle ABC$

Triangles that are congruent have three congruent angles and three congruent sides. Although this is true, you do not need to show that all three sides and angles are congruent to identify that the triangles themselves are congruent. Instead, there are four different ways to prove that any two triangles are congruent. There is also an additional test for right triangles. Before further exploring these tests, we will take a closer look at congruence.

How to Indicate Congruence

You have probably noticed before that the math symbol that signifies congruence is ≅. The statement $\angle M \cong \angle N$ is read *angle M is congruent to angle N*. The way to indicate when two items are not congruent is with a diagonal line through the congruence symbol, such as: "≇." In the following figure you can see that the angles are marked to indicate three pairs of congruent angles. You can also see that the sides are each marked to signify congruence.

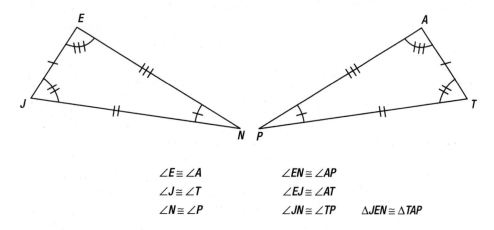

$$\angle E \cong \angle A \qquad \angle EN \cong \angle AP$$
$$\angle J \cong \angle T \qquad \angle EJ \cong \angle AT$$
$$\angle N \cong \angle P \qquad \angle JN \cong \angle TP \qquad \triangle JEN \cong \triangle TAP$$

Order Matters with Congruence

In the previous figure, we could refer to $\triangle JEN$ as $\triangle EJN$, $\triangle JNE$, or $\triangle NEJ$. Starting point and direction are not important when you're simply *naming* a triangle. However, when you're discussing congruence, how you name shapes is very important. The order of the congruent vertexes must correspond from one triangle to the other triangle. Refer to the preceding figure again for the following example.

> Since $\angle J \cong \angle T$, these two angles must be in the same position in the triangle's names when you're noting their congruence.
>
> The same goes for $\angle E$ and $\angle A$. It would be correct to write $\triangle EJN \cong \triangle ATP$ since in this ordering, congruent angles E and A are both first, while congruent angles J and T are both second.
>
> Why would it be incorrect to write $\triangle ENJ \cong \triangle PAT$?

TIP: When you are naming congruent triangles, the corresponding pairs of congruent angles must be in the same order.

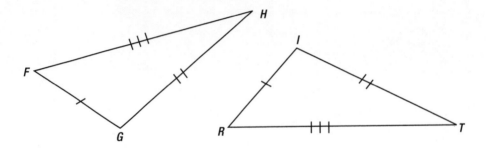

Don't Trust Appearances!

Sometimes, it might be difficult to tell that two triangles are congruent if they are oriented differently. For example, in the preceding figure, the two triangles are congruent, but it takes a little investigation to see what the corresponding parts are. An easy starting point is to find the angles that sit *between* pairs of congruent sides in each of the triangles. Since angles G and I are between two pairs of corresponding, congruent sides, this shows that $\angle G \cong \angle I$. The same is true of angles F and R. Therefore, when you're referring to the congruence of these triangles, the names will reflect these pairings: $\triangle FGH \cong \triangle RIT$. What are two other ways that you could name the two triangles to show their congruence?

TIP: When you are given information about congruent shapes, the order of the letters shows which angles and sides correspond to each other and are congruent.

PRACTICE 1

Use the following figure in questions **1–5** to identify congruent triangles and angles.

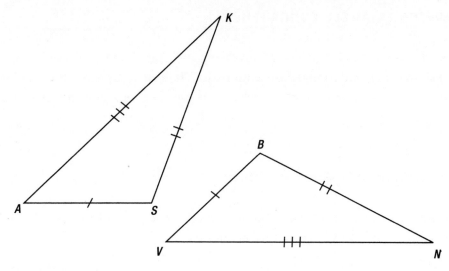

1. $\triangle ASK \cong \triangle$_____

2. $\triangle KAS \cong \triangle$_____

3. $\triangle SKA \cong \triangle$_____

4. $\angle A \cong \angle$_____

5. $\angle K \cong \angle$_____

For questions **6–10**, identify the congruent parts of the triangle, given that $\triangle WEB \cong \triangle INK$. (*Remember:* Order matters.)

6. Side $EB \cong$ Side ____

7. Side $BW \cong$ Side ____

8. Side $NI \cong$ Side ____

9. $\angle E \cong \angle$_____

10. $\triangle NKI \cong \triangle$_____

PROVING TRIANGLE CONGRUENCE

Now that you have an understanding of congruence, it is time to explore the tests that show when two triangles are congruent. There are four tests, or *postulates*, that can be used to prove that triangles are congruent, with one additional test for right triangles.

Test 1. Side-Side-Side Postulate

As seen before, if all three sides of one triangle are equal in length to all three sides of another triangle, then the triangles are congruent.

Test 2. Side-Angle-Side Postulate

If two sides and the angle that is *formed by those two sides* are congruent, then the triangles are congruent. The angle *must* be the included angle between the two congruent sides for this test to be true. Looking at the next figure, can you notice that ∠C ≅ ∠Z and that both of these angles are the included angles formed by the congruent sides?

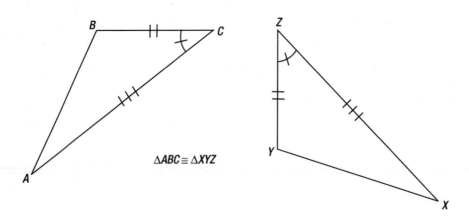

△ABC ≅ △XYZ

Test 3. Angle-Side-Angle Postulate

If two triangles have two pairs of angles and the side *included between them* is congruent, then the triangles are congruent. Looking at the next figure, can you notice that side CA is congruent to side JE? Since sides CA and EJ are between

two pairs of congruent angles, the Angle-Side-Angle test works to prove that
$\triangle CAR \cong \triangle JET$.

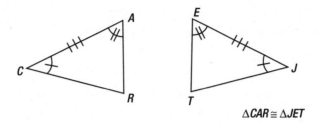

$\triangle CAR \cong \triangle JET$

Test 4. Angle-Angle-Side Postulate

If two triangles have two pairs of congruent angles and one pair of congruent
sides, then the triangles are congruent. This test is similar to Angle-Side-Angle,
except with this test the side does not have to be between pairs of congruent
angles. The next figure demonstrates this test.

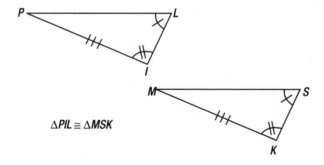

$\triangle PIL \cong \triangle MSK$

Test 5. Hypotenuse-Leg Postulate (Right Triangles Only)

This postulate applies only to right triangles. If any two right triangles have a
hypotenuse and one leg that are congruent, then the triangles are congruent. This
is true for two reasons: The first reason concerns the Pythagorean theorem, which
could be used to prove that the length of the third sides must also be congruent.
Then the Side-Side-Side postulate proves that the triangles are congruent.

Test 6. Leg-Leg Postulate (Right Triangles Only)

Similar to the previous postulate, Hypotenuse-Leg, the Leg-Leg postulate proves
congruence only for right triangles. The angle that is included between the two

legs of any right triangle is the 90-degree right angle, so this postulate is actually an abbreviation of the Side-Angle-Side postulate.

Test 7. Hypotenuse-Acute Postulate (Right Triangles Only)

The final postulate is also only applicable to proving congruence in right triangles. If the hypotenuse and an acute angle of one right triangle are congruent to the hypotenuse and the corresponding acute angle of another right triangle, then the triangles are congruent. (This is because they will also have a 90-degree angle in common, so this postulate is a short version of the Angle-Angle-Side postulate that applies to all triangles.)

· ·

TIP: The postulates that prove congruence in triangles are:

Side-Side-Side
Side-Angle-Side
Angle-Side-Angle
Angle-Angle-Side
Hypotenuse-Leg (Right Triangles Only)
Leg-Leg (Right Triangles Only)
Hypotenuse-Acute (Right Triangles Only)

· ·
· ·

TIP: Angle-Angle-Angle does NOT prove congruence in triangles. Angle-Side-Side also does NOT prove triangle congruence.

· ·

CORRESPONDING PARTS IN CONGRUENT TRIANGLES

You have now learned that when two geometric figures are congruent, their angles and sides are all congruent. You also understand that when two sides of triangles are *corresponding* it means they are in between two pairs of corresponding congruent angles. An important postulate to remember can be summed up with the acronym CPCTC. This stands for: ***Corresponding Parts of Congruent Triangles Are Congruent***. Therefore, if you know that two triangles are congruent, you also know that all of their corresponding parts are also congruent.

....

TIP: Corresponding parts of congruent triangles are congruent.

....

PRACTICE 2

1. In △ABC and △LMN, A and L are congruent, B and M are congruent, and C and N are congruent. Using the preceding information, which postulate proves that △ABC and △LMN are congruent?

2. The Springfield cheerleaders need to make three identical triangles. The girls decide to use an arm's length to separate each girl from her two other squad mates. Which postulate proves that their triangles are congruent?

Use the following figure to answer questions **3–5**.

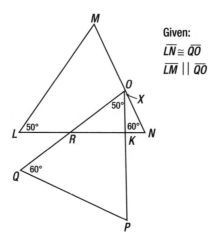

Given:
$\overline{LN} \cong \overline{QO}$
$\overline{LM} \parallel \overline{QO}$

3. Name each of the triangles in order of corresponding vertices.

4. Name corresponding line segments.

5. State the postulate that proves △LMN is congruent to △OPQ.

Use the next figure for questions 6–8.

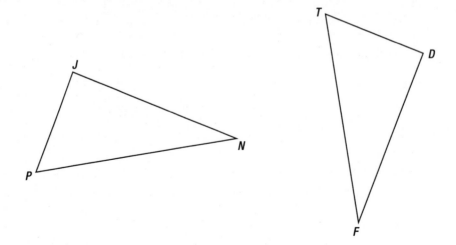

6. If given that $PJ \cong TD$ and $JN \cong DF$, what other piece of information would prove them congruent, and what postulate would be used?

7. If each triangle in this figure were a right triangle, and you were given that $JP \cong DT$, what other piece(s) of information would you need to prove congruence, and what postulate would you use?

8. Given the following information, can you determine triangle congruence? If so, using which postulate: $FT \cong NP$, $\angle N \cong \angle F$, or $DT \cong JP$?

9. Tina and Tim are making triangular trivets in woodworking class. Their teacher gives each of them three pieces of wood that measure 8 inches, 9.5 inches, and 13 inches in length. Can Tina and Tim make two different trivets or must their triangles be congruent?

ANSWERS

Practice 1

1. $\triangle ASK \cong \triangle VBN$
2. $\triangle KAS \cong \triangle NVB$
3. $\triangle SKA \cong \triangle BNV$
4. $\angle A \cong \angle V$
5. $\angle K \cong \angle N$
6. Side $EB \cong$ Side NK
7. Side $BW \cong$ Side KI
8. Side $NI \cong$ Side EW
9. $\angle E \cong -N$
10. $\triangle NKI \cong \triangle EBW$

Practice 2

1. There is no postulate Angle-Angle-Angle that proves triangle congruence, so we cannot be certain that $\triangle ABC$ and $\triangle LMN$ are congruent.
2. Side-Side-Side
3. $\triangle OQP \cong \triangle LNM$ and $\triangle QOP \cong \triangle NLM$ are two correct examples.
4. $OQ \cong LN$, $QP \cong NM$, and $PO \cong ML$
5. Angle-Side-Angle
6. Side-Side-Side could be used if you know that $PN \cong TF$. Or: Side-Angle-Side could be used if you know that $\angle J \cong \angle D$.
7. Hypotenuse-Leg would prove congruence if $PN \cong TF$
8. You cannot use Side-Angle-Side since the angle is not included *between* the two pairs of congruent sides. Since there is no postulate for Angle-Side-Side, you cannot prove congruence in this case.
9. If the three sides of a triangle are congruent, then the triangles must be congruent, so Tina and Tim will make the same trivets.

polygons

The value of a problem is not so much coming up with the answer as in the ideas and attempted ideas it forces on the would-be solver.

—I. N. HERSTEIN

In this lesson we will learn about classifying polygons and working with their angles.

WHAT ARE POLYGONS?

Polygons are closed geometric figures that are made up with three or more straight line segments as sides. In the previous lesson we discussed triangles, which are the most basic types of polygons. Polygons have *interior angles* at each of their vertices. They also have *diagonals,* which are line segments that connect any two nonadjacent vertices. Polygons are named by listing their vertices in clockwise or counterclockwise order. It does not matter which vertex is used as the starting point as long as the remaining vertices are listed in order. The following figure illustrates a six-sided hexagon that could be correctly named *ABCDEF* or *CBAFED*. It could have many other names that start with a different vertex, and the vertices could be named in clockwise or counterclockwise order.

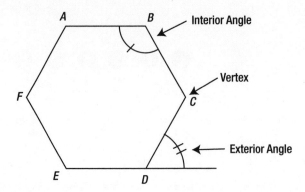

CLASSIFYING POLYGONS

The most basic way to classify a polygon is by the number of sides it has. They are further classified by whether their angles and/or sides are equal in measure. Next is a list of the most common polygons according to number of sides. You should memorize these:

Number of Sides	Polygon Name
3	Triangle
4	Quadrilateral
5	Pentagon
6	Hexagon
7	Heptagon
8	Octagon
9	Nonagon
10	Decagon

Polygons are further classified by the congruence of their sides or angles. An **equilateral** polygon has *sides* that are all equal in length. An **equiangular** polygon has *angles* that are all equal in measure. The most specific polygon is the **regular** polygon, which is both equilateral *and* equiangular. An equilateral triangle is an example of a three-sided regular polygon. A rectangle is equiangular, but since it does not have to be equilateral, it is not a regular polygon. However, a square *is* regular since its sides and angles are *all* congruent. You can notice the difference in the next figure.

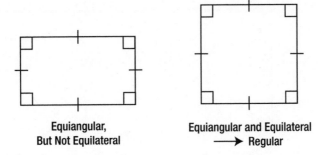

Equiangular,
But Not Equilateral

Equiangular and Equilateral
⟶ Regular

PRACTICE 1

1. A polygon that has all congruent sides is _____.

2. A polygon that has all congruent sides *and* all congruent angles is

_____.

3. True or false: Polygons must be named by listing their vertices in a *clockwise* fashion.

4. A polygon with eight sides is called a/an _____.

5. True or false: A polygon that has eight congruent sides must also have eight congruent angles.

ANGLES OF A POLYGON

Interior Angles

When we are discussing the angles of a polygon, we consider the sum of the interior and exterior angles. You already know that the sum of the interior angles of a triangle is 180°. For every additional side that polygons have, their interior angle sum increases by 180°. This is because for every side above three sides, an additional triangle can be formed inside the polygon by connecting the vertices. This is illustrated in the next figure.

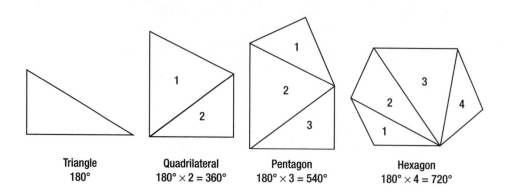

Triangle	Quadrilateral	Pentagon	Hexagon
180°	180° × 2 = 360°	180° × 3 = 540°	180° × 4 = 720°

In this figure you can see how a quadrilateral contains two triangles, a pentagon contains three triangles, and a hexagon contains four triangles. Multiplying the number of triangles within each polygon will result in the sum of the interior angles of the geometric figure. The formula for calculating the sum of the interior angles of a polygon with n sides is $180°(n-2)$. If you forget this formula, start with a single vertex and draw nonoverlapped triangles with diagonals to the other vertices. Then, multiply the number of nonoverlapping triangles it takes to fill the polygon by 180°.

TIP: The sum of the measures of the interior angles of any polygon with n sides is $180°(n-2)$.

In the event of **equiangular polygons** and **regular polygons**, the measure of one of the interior angles is found by dividing the sum of the interior angles by the number of vertices or sides. For example, a square is a regular quadrilateral. Since regular quadrilaterals have an interior angle sum of 360°, each angle has a measure of $\frac{360°}{4}$ or 90°. Although a five-sided regular pentagon will have the same interior angle sum as an irregular pentagon, you can identify the individual interior angle measures only in *regular* polygons since they are equiangular.

TIP: In a regular polygon with n sides, the measure of an interior angle is $180°\left(\frac{n-2}{n}\right)$.

Exterior Angles

You probably remember learning about exterior angles in the previous lessons on triangles. Exterior angels exist between a polygon's side and the extension of its adjacent side. Regardless of how many sides a polygon has, the sum of its exterior angles will always be 360°. Therefore, in the case of a regular polygon with n sides, the measure of an exterior angle will be $\frac{360°}{n}$.

....

TIP: In a regular polygon with n sides, the measure of an exterior angle is $\frac{360°}{n}$.

....

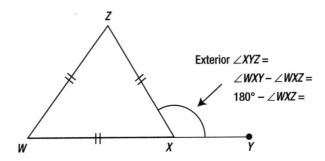

Exterior $\angle XYZ =$
$\angle WXY - \angle WXZ =$
$180° - \angle WXZ =$

Can you see in this figure that exterior angle YXZ of the regular triangle is the supplement of its interior angle WXZ? Together, they make a straight angle and sum to 180°. Therefore, for regular polygons, an alternative way to identify the measure of an interior angle is by subtracting the measure of the exterior angle by 180°.

....

TIP: In all regular polygons, the measure of an interior angle will equal 180° − (measure of an exterior angle).

....

PRACTICE 2

1. True or false: The sum of the measures of the exterior angles of a regular polygon is identified by how many sides the polygon has.

2. What is the measure of an exterior angle of an equilateral triangle?

3. A polygon has 12 sides. What is the sum of the measures of its interior angles?

4. What is the measure of an interior angle of a regular nonagon?

5. If the sum of the measures of a polygon's interior angles is 2,700°, determine how many sides that polygon must have.

6. Determine the number of sides a regular polygon must have if one of its interior angles has a measure of 168°.

7. A porcelain tile that is a regular hexagon is used to cover the floor in the foyer of a hotel. When one of the tiles is pressed against a wall so that its side is evenly against the wall, what will be the angle between the wall and the adjacent side (see next figure)?

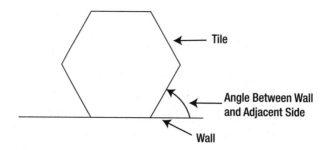

8. Draw an irregular decagon. From one vertex, create as many nonoverlapping triangles as possible. Draw diagonals to the nonadjacent vertices. How many non-overlapping triangles can you create?

9. Use your illustration from question number 8 to identify the sum of the interior angles of a decagon.

ANSWERS

Practice 1

1. equilateral
2. regular
3. false
4. octagon
5. False. A polygon with eight congruent sides is equilateral, but does not have to be equiangular.

Practice 2

1. False. The sum of the measures of the exterior angles of any polygon is always 360°.
2. 120° (180° − 60 = 120°)
3. 1,800° (Using the formula 180°[$n-2$], we get 180°[12 − 2] = 1,800°)
4. 140° (Using the formula $180°\frac{n-2}{n}$, we get $180°\frac{9-2}{9} = 140°$)
5. The polygon must have 17 sides.

     180°($n-2$) = 2,700

     $n - 2 = 15$, so $n = 17$
6. $n = 30$

     $180°\frac{n-2}{n} = 168°$

     $180°n - 360 = 168n$

     $-360 = -12n$, so $n = 30$
7. 60° ($\frac{360°}{6} = 60°$)
8. eight non-overlapping triangles
9. 1,440°

quadrilaterals part i: parallelograms

Without geometry, life is pointless.
—UNKNOWN

In this lesson you will learn how to classify quadrilaterals according to their parallel sides, and you will learn the main properties of parallelograms.

CLASSIFYING QUADRILATERALS

Quadrilaterals are closed geometric figures that are made up from four sides of straight-line segments. As we discussed in the previous lesson, a four-sided polygon contains two non-overlapping triangles inside it, and it has an interior angle sum of 360°. This is true of all quadrilaterals regardless of their appearance or congruent parts. Quadrilaterals are divided into three major categories, according to how many pairs of opposite parallel sides they have.

1. The first group has no pairs of parallel sides. It may or may not have sides and angles that are congruent, but it does not have any parallel sides.

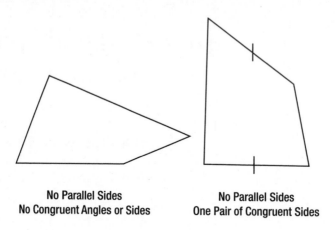

No Parallel Sides
No Congruent Angles or Sides   **No Parallel Sides**
One Pair of Congruent Sides

2. The second group has exactly one pair of parallel sides and is referred to as **trapezoids**. If a trapezoid has one pair of opposite congruent sides, then it is an **isosceles trapezoid**. See the next figure.

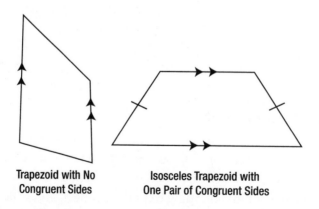

Trapezoid with No   **Isosceles Trapezoid with**
Congruent Sides   **One Pair of Congruent Sides**

3. The third classification of quadrilaterals is **parallelograms**. Parallelograms have two pairs of parallel sides.

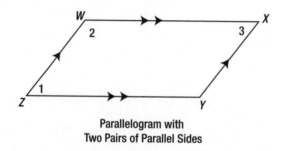

Parallelogram with
Two Pairs of Parallel Sides

PROPERTIES OF PARALLELOGRAMS

Parallelograms are further classified by the number of congruent angles and sides they have. In this lesson we are going to begin by learning the basic properties that apply to all parallelograms.

Interior Angles

In Lesson 5 we discussed the properties of congruent lines that are intersected by a transversal. In the parallelogram in the preceding figure, note that side WZ is a transversal through parallel sides WX and ZY. Angle 1 and angle 2 can then be considered *same-side interior* angles and one of our tests from Lesson 5 states that same-side interior angles are always supplementary. Therefore, we can conclude that in parallelograms, adjacent pairs of angles are supplementary.

TIP: Adjacent pairs of angles in parallelograms are supplementary.

Continuing our study of the figure with two pairs of parallel sides, we see that $\angle 1 + \angle 2 = 180°$ and that $\angle 3 + \angle 2 = 180°$. From there we can next identify that $\angle 1 + \angle 2 = \angle 3 + \angle 2$, since both of these sums equaled 180°. If we subtract $\angle 2$ from each side, we see that $\angle 1 = \angle 3$. This brings us to the next important property of parallelograms, which is that opposite pairs of angles are congruent. It is also true that any quadrilateral that has two pairs of opposite congruent angles must be a parallelogram.

TIP: Opposite pairs of angles in parallelograms are congruent.

These two properties can be seen in the next figure.

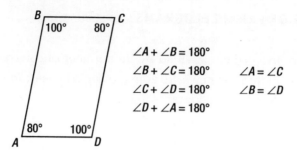

$$\angle A + \angle B = 180°$$
$$\angle B + \angle C = 180° \qquad \angle A = \angle C$$
$$\angle C + \angle D = 180° \qquad \angle B = \angle D$$
$$\angle D + \angle A = 180°$$

PRACTICE 1

1. A quadrilateral with one pair of parallel sides is called a/an

   _____.

2. True or false: A quadrilateral with one pair of parallel sides and one pair of congruent sides is called an equilateral trapezoid.

3. In quadrilateral *ASDF*, $m\angle A = 117°$, $m\angle S = 39°$, and $m\angle D = 86°$. Find the $m\angle F$.

4. In parallelogram *RTYU*, $m\angle R = 74°$. Find the measures of the other three angles.

5. In the next figure, $m\angle P$ is twice as large as $m\angle R$. Use the given information to determine the measures of angles *P* and *R*.

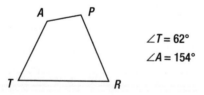

$$\angle T = 62°$$
$$\angle A = 154°$$

6. In quadrilateral *HJKL*, the measures of angles *H*, *J*, *K*, and *L* are in a ratio of 1:2:3:4. Find the measure of each of the angles in the quadrilateral.

7. True or false: In a parallelogram, there is one pair of congruent opposite angles only when there is one pair of congruent sides.

Use the next figure to answer questions **8–11**. The figures *are not drawn to scale*. Only the markings on the sides and angles should be used to identify the *most specific name* you can give to each four-sided polygon.

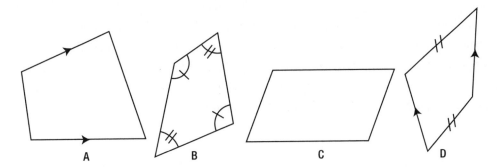

8. Shape *A* is a _____.

9. Shape *B* is a _____.

10. Shape *C* is a _____.

11. Shape *D* is a _____.

Diagonals and Opposite Sides

Look at the diagonal drawn into parallelogram *RTGB* in the next figure. Notice that angles 1 and 4 are alternate interior angles formed by parallel line segments *RB* and *TB* and by transversal *BT*. From Lesson 5, we know that alternate interior angles are congruent, so we can conclude that $m\angle 1 = m\angle 4$. The same is true of angles 2 and 3, which are alternate interior angles between line segments *RT* and *BG*: $m\angle 2 = m\angle 3$. Since segment *BT* = *BT*, we can conclude that $\triangle RBT \cong \triangle GTB$ by the postulate Angle-Side-Angle. This is our next test of parallelograms: Their diagonals divide them into two congruent triangles.

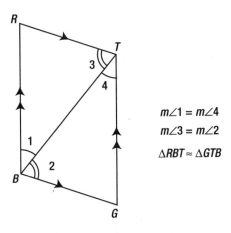

$m\angle 1 = m\angle 4$
$m\angle 3 = m\angle 2$
$\triangle RBT \approx \triangle GTB$

. . . .

TIP: A parallelogram's diagonal divides it into two congruent triangles.

. . . .

Once it is known that a diagonal divides a parallelogram into two congruent triangles, it can also be stated that the opposite sides of parallelograms are congruent. This is true because the opposite sides of the parallelogram are corresponding parts of the triangles formed by the diagonal. For example, in the preceding figure, side *RT* corresponds with side *BG*.

. . . .

TIP: Parallelograms contain two pairs of opposite, congruent sides.

. . . .

The last property of parallelograms that we are going to cover in this lesson is one more detail about diagonals. The two diagonals of any parallelogram will bisect each other. You probably remember from earlier lessons that *bisect* means *to cut into two congruent parts.*

. . . .

TIP: A parallelogram's diagonals bisect each other.

. . . .

HOW TO PROVE THAT A QUADRILATERAL IS A PARALLELOGRAM

You just learned several properties that are true when a quadrilateral is a parallelogram. Sometimes, you will be given information about a quadrilateral, and you will want to know if that means that the quadrilateral is a parallelogram. There are five ways that you can be certain that a quadrilateral is a parallelogram:

- 2 pairs of opposite, parallel sides
- 2 pairs of opposite, congruent sides
- 2 pairs of opposite, congruent angles
- diagonals that bisect each other
- 1 pair of opposite sides that are parallel *and* congruent

PRACTICE 2

1. What two facts are always true about the diagonals of a parallelogram?

2. In quadrilateral *WEBS*, *WE* ∥ *BS*, and *WS* is not congruent to *EB*. What kind of quadrilateral is *WEBS*?

3. NEWS is a parallelogram. Side *WE* = 4x – 10 and side *SN* = 3x + 9. Find the length of side *WE*.

Use the next figure to answer questions **4** and **5**.

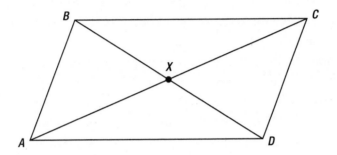

4. If *ABCD* is a parallelogram with *BX* = 4x – 3 and *DX* = 21 – 2x, find the length of *BD*.

5. If *ABCD* is a parallelogram with *m∠ABC* = 5y – 10 and *m∠BCD* = 3y – 18, find the value of *y* and the *m∠BAD*.

6. Find the value of *q* that would prove that *ABCD* is a parallelogram given the following information: *m∠BDC* = 80°, *m∠BCA* = 30°, *m∠BDA* = (2q)°, and *m∠ACD* = (4q + 4)°.

7. In parallelogram *QPIE*, *m∠P* = (3u – 8)° and *m∠E* = (127 – 2u)°. Find the measures of all of the angles of *QPIE*.

ANSWERS

Practice 1

1. trapezoid
2. False. It is called an isosceles trapezoid.
3. $m\angle F = 118°$
4. $m\angle T = 106°$, $m\angle Y = 74°$, and $m\angle U = 106°$. (In a parallelogram, adjacent angles are supplementary, and opposite angles are congruent.)
5. $m\angle P = 96$ and $m\angle R = 48$. (Since $360 - 62 - 154 = 144$, angles P and R must sum to 144. Since $m\angle P$ is twice as big as $m\angle R$, set up and solve the equations $2x + x = 144$, where $m\angle P = 2x$ and $m\angle R = x$.)
6. $36°, 72°, 108°$, and $144°$. (With ratio problems, multiply each number by the same variable x and set up an equation: $1x + 2x + 3x + 4x = 360$, so $10x = 360$ and $x = 36$.)
7. False. Parallelograms always have two pairs of congruent opposite angles.
8. Trapezoids have exactly one pair of parallel sides.
9. Parallelograms have two pairs of opposite, congruent angles.
10. The polygon *looks* like a parallelogram, but it has no markings to indicate congruence or parallel sides, so it can only be classified as a quadrilateral.
11. Isosceles trapezoids have one pair of parallel sides and one pair of opposite congruent sides.

Practice 2

1. First fact: Any diagonal of a parallelogram will divide it into two congruent triangles.
   Second fact: The two diagonals of any parallelogram bisect each other.
2. *WEBS* is a trapezoid since it has one pair of parallel sides. It cannot be a parallelogram, since opposite sides are congruent in parallelograms.
3. $WE = 66$. Opposite sides are congruent in parallelograms: Solve $4x - 10 = 3x + 9$ to get $x = 19$, so $WE = 66$.
4. $BD = 26$. Since diagonals bisect each other in parallelograms, $4x - 3 = 21 - 2x$, and $x = 4$. Therefore, $BX = BD = 13$.
5. $m\angle BAD = 60°$. Since adjacent angles in parallelograms are supplementary, $(5y - 10) + (3y - 18) = 180°$. Therefore, $y = 26$, and since $m\angle BAD = m\angle BCD = 3y - 18$, then $m\angle BAD = 60°$.
6. $q = 11$. Adjacent angles are supplementary in parallelograms, so the following equation can be used: $m\angle BDC + m\angle BCA + m\angle BDA + m\angle ACD = 180°$. Using substitution, we get $80° + 30° + (2q)° + (4q + 4)° = 180°$ and solved we see that $q = 11$.
7. $m\angle P = m\angle E = 73°$ and $m\angle Q = m\angle I = 107°$. Opposite angles are congruent in parallelograms so $(3u - 8) = (127 - 2u)$ and $u = 27$.

quadrilaterals part ii: trapezoids and special parallelograms

Life is good for only two things, discovering mathematics and teaching mathematics.
—SIMÉON POISSON

In this lesson you will expand your understanding of the quadrilateral family. You will learn about trapezoids and special parallelograms such as squares, rectangles, and rhombuses.

THE QUADRILATERAL FAMILY TREE

You already know that quadrilaterals are four-sided figures, which are classified by the number of parallel sides they have. The most basic quadrilateral is a four-sided polygon, which has no congruent angles or sides. Starting from here, a family tree can be drawn to give a visual picture of how the different quadrilaterals relate to one another. Notice in the Quadrilateral Family Tree in the following figure, the further down the tree you go, the more specific the classification is.

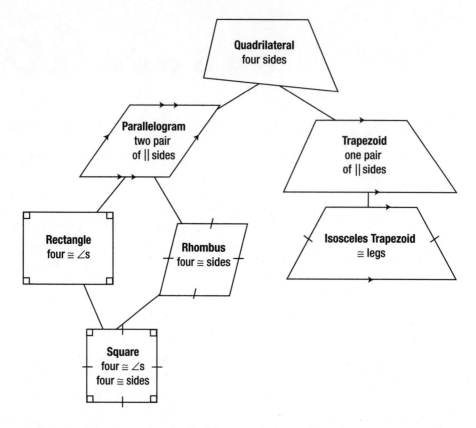

Trapezoids are on one-half of the quadrilateral family tree. Trapezoids are quadrilaterals with exactly one pair of parallel sides. A special type of trapezoid is the **isosceles trapezoid**, which has one pair of congruent sides.

The other half of the quadrilateral family tree represents parallelograms, which as you remember from the last lesson, are four-sided polygons that have two pairs of parallel sides. Although we discussed the special properties of parallelograms in the previous lesson, we did not investigate **rectangles**, **rhombuses**, or **squares**. These shapes are unique parallelograms with congruent angles, congruent sides, or congruent angles *and* sides:

TIP: A *rectangle* is a parallelogram with four congruent angles. Each angle measures 90 degrees.

A *rhombus* is a parallelogram with four congruent sides. (The plural of *rhombus* is either *rhombuses* or *rhombi*.)

A *square* is a parallelogram with four congruent angles *and* four congruent sides. Squares are both rectangles and rhombuses.

PRACTICE 1

Use the family tree diagram to identify whether each statement is true or false.

1. All trapezoids are quadrilaterals. _____

2. All parallelograms are rectangles. _____

3. All rectangles are quadrilaterals. _____

4. All parallelograms are squares. _____

5. All squares are rhombuses. _____

6. All rectangles are parallelograms. _____

7. All rhombuses are squares. _____

8. All isosceles trapezoids are quadrilaterals. _____

9. All squares are trapezoids. _____

10. All quadrilaterals are squares. _____

OTHER SPECIAL PROPERTIES

In the previous lesson you learned that the diagonals of parallelograms always bisect each other. Now that you know how the different types of parallelograms relate to each other, here are a few special properties of rectangles, rhombuses, and squares to learn.

....

TIP: The diagonals of a rectangle are always congruent.

....

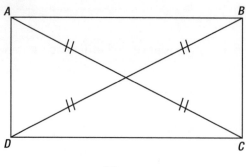

$$\overline{AC} = 16$$
$$\overline{DB} = 16$$

You can see in the preceding figure that diagonals *AC* and *DB* are congruent.

....

TIP: The diagonals of a rhombus are always perpendicular bisectors.

....

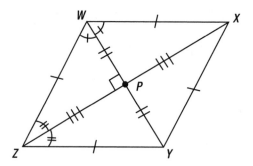

You can see in this figure that diagonals *WY* and *XZ* form four right angles. Although diagonal *XZ* is longer than diagonal *WY*, they still bisect each other.

....

TIP: The diagonals of a rhombus bisect the angles at each vertex of the rhombus.

....

Looking again at this figure, notice that ∠ZWY is congruent to ∠XWY. Also, ∠WZX ≅ ∠YZX.

Since a square has the properties of a rhombus *and* a rectangle, squares have the properties of both of these shapes.

TIP: The diagonals of a square are congruent, perpendicular bisectors that bisect the angles of the square.

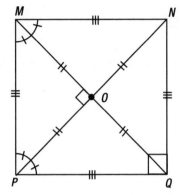

This figure illustrates all of the special properties of squares. What type of special triangle is $\triangle MOP$? This is an isosceles right triangle.

PRACTICE 2

Use the following figure and the given information to answer questions **1–6**.

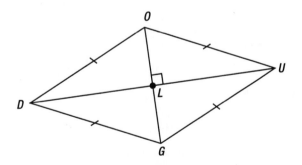

Given: $m\angle ODL = 43°$
          $\overline{DU} = 16$
          $\overline{DO} = 10$

1. What type of quadrilateral is $DOUG$?

2. What is the $m\angle LDG$?

3. What is the $m\angle DOU$?

4. What is the $m\angle LGU$?

5. What is the length of $\overline{LU}$?

6. What is the length of $\overline{LG}$?

Use the rectangle and the given information in the following figure to answer questions **7–10**.

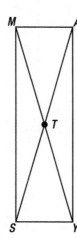

Given: $m\angle ATY = 142°$
         $\overline{ST} = 17.6$

7. What is the length of $\overline{MY}$?

8. What is the $m\angle TAY$?

9. What is the $m\angle TMA$?

10. If the length of $\overline{MY}$ is $2f + 13.2$, find the value of f.

MORE FUN WITH TRAPEZOIDS

In a trapezoid, the pair of parallel sides are referred to as the **bases** of the trapezoid and the non-parallel sides are its **legs**. In the next figure these parts are labeled. Another important term relating to trapezoids is **altitude**. You are already familiar with the altitude of a triangle: It is the height from one vertex to the opposite side, drawn at a right angle. In a trapezoid, the altitude is a perpendicular segment extending from one of the parallel bases to the other parallel base. Notice that in this figure, trapezoids can have more than one altitude.

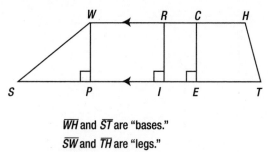

$\overline{WH}$ and $\overline{ST}$ are "bases."
$\overline{SW}$ and $\overline{TH}$ are "legs."
$\overline{WP}$, $\overline{RI}$, and $\overline{CE}$ are all "altitudes."

Median is another important term used with trapezoids. The word "median" sounds like "medium." When something is "medium," it is in the middle—it is not too big and not too small. The *median* of a trapezoid is the line segment that joins the midpoints of the legs. The median is parallel to both bases and its length is the *average of the bases*. Take a look at the next figure in order to gain a visual understanding of the median. If base *ST* measured 10 inches and base *KF* measured 14 inches, then median *IA* would be 12 inches, since 12 is the average of 10 and 14.

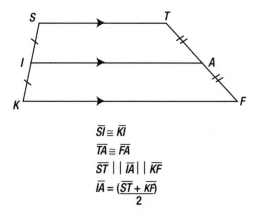

$\overline{SI} \cong \overline{KI}$
$\overline{TA} \cong \overline{FA}$
$\overline{ST} \mid\mid \overline{IA} \mid\mid \overline{KF}$
$\overline{IA} = \dfrac{(\overline{ST} + \overline{KF})}{2}$

....

TIP: The *median* of a trapezoid is the segment that connects the midpoints of its non-parallel legs. The length of the median is the average of the two bases.

....

PRACTICE 3

Use this figure to answer questions **1–5**.

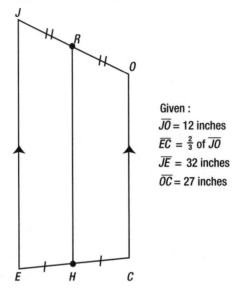

Given :
$\overline{JO} = 12$ inches
$\overline{EC} = \frac{2}{3}$ of $\overline{JO}$
$\overline{JE} = 32$ inches
$\overline{OC} = 27$ inches

1. What is the length of $\overline{RO}$?

2. What is the length of $\overline{EH}$?

3. Segment $\overline{RH}$ is called the _____ of the trapezoid.

4. What is the length of $\overline{RH}$?

5. If $\overline{OC} = 2y + 8$ and $\overline{JE} = 3y - 6$, write an expression in terms of y to represent the length of $\overline{RH}$.

ANSWERS

Practice 1

1. True
2. False
3. True
4. False
5. True
6. True
7. False
8. True
9. False
10. False

Practice 2

1. *DOUG* is a rhombus since it is a quadrilateral with four congruent sides.
2. $m\angle LDG = 43°$
3. $m\angle DOU = 94°$
4. $m\angle LGU = 47°$
5. $\overline{LU} = 8$
6. $\overline{LG} = 6$ (ΔLGU is a right triangle, so use the Pythagorean theorem or use your knowledge of 3-4-5 triangles to solve for $\overline{LG}$.)
7. $\overline{MY}$ is 35.2, twice the length of $\overline{ST}$.
8. $m\angle TAY = 19°$ (ΔTAY is isosceles and $\angle TAY$ is one of the congruent base angles)
9. $m\angle TMA = 71°$ ($m\angle MTA = 38°$, so there are 142° split evenly between $\angle TMA$ and $\angle TAM$.)
10. $\overline{MY} = 2(\overline{ST})$
    $2f + 13.2 = 2(17.6)$
    $2f + 13.2 = 35.2$
    $2f = 22$
    $f = 11$

Practice 3

1. $\overline{RO} = 6$ inches
2. $\overline{EH} = 4$ inches. ($\frac{2}{3}$ of 12 is 8 inches, so $\overline{EC} = 8$ inches and $\overline{EH}$ is half of that.)
3. median
4. $\overline{RH} = 29.5$ inches. (The median is the average of the two bases: $\frac{32 + 27}{2} = 29.5$ inches.)
5. $\overline{RH} = 2.5y + 1$. Take the average of both bases: $\frac{(2y+8)+(3y-6)}{2} = \frac{5y+2}{2} = 2.5y + 1$.

LESSON 17

ratios and proportions

Do not worry too much about your difficulties in mathematics;
I can assure you that mine are still greater.
—ALBERT EINSTEIN

In this lesson you will explore ratios and proportions. You will learn how to set up proportions to solve problems, which will be applied to similar and congruent triangles in the next lesson.

RATIOS AND PROPORTIONS are widely used concepts in the everyday world. A common example of a ratio that you might see in a supermarket is "3 boxes of cereal for the price of 2." A common use of proportions is estimating distances between cities when working with the scale on a map. Before we get into problem solving with ratios and proportions, you first need to understand what ratios are and how they are used to construct proportions.

RATIOS: THE BUILDING BLOCK OF PROPORTIONS

A **ratio** is a comparison of two different quantities or numbers. Ratios can be written with colons or as fractions. They have the same meaning, regardless of which style is used to present them. The following two ratios are said "dogs to cats": dogs:cats and $\frac{\text{dogs}}{\text{cats}}$. Notice that the first word said must be the first term when you're using a colon or it must be the numerator of the fraction.

> **TIP:** A *ratio* is a comparison of two numbers, written as the quotient of two quantities in fraction form, or as two terms separated by a colon.

Ratios are always reduced into simplest terms. For example, if an animal shelter has 20 dogs and 30 cats, the ratio of dogs to cats will be expressed in simplest terms by reducing $\frac{20}{30}$ to $\frac{2}{3}$. This can also be written as 2:3. Ratios sometimes represent the "part" of a quantity to the "whole" of the quantity. In the preceding case of the animal shelter, there are 50 animals total (20 dogs and 30 cats), so the ratio of dogs to animals would be $\frac{20}{50}$ or $\frac{2}{5}$. One important thing to keep in mind when you're reducing ratios to simplest terms, is that they should always reflect two quantities and not be written as just a whole number. For example, if the number of apples to oranges in a fruit basket is 12 to 4, this is written as $\frac{12}{4} = \frac{3}{1}$. It is not written as just "3."

PRACTICE 1

1. What is the ratio of weekdays to weekend days in any given week?

2. What is the ratio of weekdays to the total number of days in a week?

Consider the set of whole numbers 1 through 10 for questions **3–5**.

3. What is the ratio of odd numbers to even numbers in this set?

4. What is the ratio of prime numbers to composite numbers in this set?

5. What could the ratio 1:2 represent, given the set of whole numbers 1–10?

Using the following figure for questions **6–9**, represent each ratio in simplest form.

6. $\frac{AB}{BC}$

7. $\frac{AB}{CD}$

8. $\frac{CD}{AB}$

9. $\frac{AC}{AD}$

PROPORTIONS: THE RELATIONSHIP BETWEEN TWO RATIOS

A **proportion** is an equation where two ratios equal each other. For example, consider the following lemonade instructions. The directions on a jar of lemonade concentrate suggest that one gallon of lemonade can be made by mixing 1 quart of lemonade concentrate with 3 quarts of water. It also suggests that for two gallons of lemonade, 2 quarts of lemonade concentrate should be mixed with 6 quarts of water. A proportion modeling the two different recipes will look like this:

$$\frac{\text{concentrate}}{\text{water}} = \frac{1}{3} = \frac{2}{6}$$

or, it would also be correct to model the directions in this manner:

$$\frac{\text{water}}{\text{concentrate}} = \frac{1}{3} = \frac{6}{2}$$

Notice that it does not matter if the lemonade concentrate is in the numerator or the denominator of the ratios. As long as a single proportion has the quantities of lemonade concentrate in the same position in both of the ratios, it is correct.

> **TIP:** A *proportion* is an equation with two ratios that are equal to each other.

THE MEANS–EXTREMES PROPERTY OF PROPORTIONS

The first and last numbers in a proportion are called the *extremes*. The middle numbers are called the *means*.

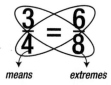

$$\frac{3}{4} = \frac{6}{8}$$

means extremes

4:6 = 2:3
means
extremes

There is a special relationship between the means and the extremes of proportions. Consider the following proportion:

$$\frac{3}{6} = \frac{1}{2}$$

You probably know that both of these fractions are equal to each other since $\frac{3}{6}$ reduces to $\frac{1}{2}$. Take a closer look at what is true when the means are multiplied together ($2 \times 3 = 6$) and when the extremes are multiplied together ($1 \times 6 = 6$). You can see that in this proportion, the product of the means equals the product of the extremes. This rule actually holds true for *all* proportions and is an important postulate when working with proportions.

If $\frac{j}{k} = \frac{e}{f}$, then $jf = ke$

If $j{:}k = e{:}f$, then $jf = ke$

....

TIP: In a *proportion*, the product of the *means* equals the product of the *extremes*. If $\frac{d}{l} = \frac{m}{s}$, then $ds = lm$.

....

Since you now know that in a proportion, the products of the means and extremes are equal, it will probably not surprise you to learn a converse truth that relates to proportions: If the product of the means does NOT equal the product of the extremes, then the proportion is not true. For example, if you wanted to test to see if the following relationship were true, you would compare the products of the means and extremes: $\frac{5}{7} = \frac{17}{20}$. In this case, the product of the means is $7 \times 17 = 119$ and the product of the extremes is $5 \times 20 = 100$. Since these two products are NOT equal, the proportion $\frac{5}{7} = \frac{17}{20}$ is NOT true. $\frac{5}{7}$ is not equivalent to $\frac{17}{20}$.

PRACTICE 2

For questions **1–3**, identify whether the given proportion is true or false.

1. $\frac{8}{9} = \frac{11}{12}$

2. $\frac{2}{5} = \frac{16}{40}$

3. $\frac{2x}{x} = \frac{2yz}{yz}$

Use the following information for questions **4–6**: A map of New York City uses a scale that represents 500 meters with $\frac{1}{2}$ inch.

4. Write a ratio that shows the relationship between meters and inches.

5. Two subway stations on the New York City map are $3\frac{1}{2}$ inches apart. Write a proportion that sets your ratio from question **4** equal to the distance in *m* meters between the two subway stations.

6. Two parks in New York City are 3,200 meters away from each other. What proportion would be used to calculate how many inches, *i*, the parks should be apart on the park map?

SOLVING PROPORTIONS

Now that you know how to set up proportions, we will take a look at solving them. Since the product of the means always equals the product of the extremes, it is easy to make an algebraic equation to solve. To demonstrate this, we will use the proportion from question **6** in the previous practice set. Beginning with the proportion, write the product of the means as equal to the product of the extremes. Then solve for the variable *i*.

$$\frac{\text{meters}}{\text{inches}} = \frac{500}{\frac{1}{2}} = \frac{3,200}{i}$$

$$500(i) = \tfrac{1}{2}(3,200)$$

$$500i = 1,600$$

$$i = \frac{1,600}{500}$$

$$i = 3\tfrac{1}{5}$$

Therefore, the two parks should be $3\frac{1}{5}$ inches apart on the map.

PRACTICE 3

Use the following information to solve questions **1–4**: In Mr. Mallory's class the ratio of girls to boys is 3:4.

  1. If there are 12 boys in the class, what proportion would be used to solve for how many girls, *g*, are in Mr. Mallory's class?

  2. If there are 12 girls in Mr. Mallory's class, how many boys are in his class?

  3. There are 21 students in Mr. Mallory's class. What proportion would be used to solve for how many boys, *b*, are in the class? (*Hint*: You will be comparing the number of boys to the total number of students.)

  4. How many girls are in Mr. Mallory's class, if there are 28 students in his class?

ANSWERS

Practice 1

1. $\frac{\text{weekdays}}{\text{weekend}} = \frac{5}{2}$

2. $\frac{\text{weekdays}}{\text{total}} = \frac{5}{7}$

3. $\frac{5}{5} = \frac{1}{1}$ (*Remember*: This cannot be written as just *1*—keep it in fraction or colon form.)

4. $\frac{\text{prime}}{\text{composite}} = \frac{4}{5}$ (*Remember*: 1 is neither prime nor composite. 2, 3, 5, and 7 are prime. 4, 6, 8, 9, and 10 are composite.)

5. The ratio 1:2 could be representing the number of even (or odd) numbers in the set to the total amount of numbers in the set, since 5:10 reduces to 1:2.

6. $\frac{AB}{AC} = \frac{5}{3}$

7. $\frac{AB}{CD} = \frac{5}{10} = \frac{1}{2}$

8. $\frac{CD}{AB} = \frac{10}{5} = \frac{2}{1}$

9. $\frac{AC}{AD} = \frac{8}{18} = \frac{4}{9}$

Practice 2

1. False. Since the product of the means does not equal the product of the extremes, this proportion is false ($9 \times 11 = 99$ and $8 \times 12 = 96$)

2. True. The product of the means equals the product of the extremes ($5 \times 16 = 80$ and $2 \times 40 = 80$)

3. True. The product of the means equals the product of the extremes ($x \times 2yz = 2xyz$ and $2x \times yz = 2xyz$)

4. $\frac{\text{meters}}{\text{inches}} = \frac{500}{\frac{1}{2}}$

5. $\frac{\text{meters}}{\text{inches}} = \frac{500}{\frac{1}{2}} = \frac{m}{3\frac{1}{2}}$

6. $\frac{\text{meters}}{\text{inches}} = \frac{500}{\frac{1}{2}} = \frac{3{,}200}{i}$

Practice 3

1. $\frac{\text{girls}}{\text{boys}} = \frac{3}{4} = \frac{8}{12}$

2. 16 boys

$\frac{\text{girls}}{\text{boys}} = \frac{3}{4} = \frac{12}{b}$

$$4(12) = 3(b)$$
$$48 = 3b$$
$$b = 16$$

3. $\frac{\text{boys}}{\text{total}} = \frac{4}{3+4} = \frac{b}{21}$

$\frac{\text{boys}}{\text{total}} = \frac{4}{7} = \frac{b}{21}$

4. There are 12 girls.

$\frac{\text{girls}}{\text{total}} = \frac{3}{3+4} = \frac{g}{28}$

$\frac{\text{girls}}{\text{total}} = \frac{3}{7} = \frac{g}{28}$

$$7(g) = 3(28)$$
$$7g = 84$$
$$g = 12$$

applying proportions to similar polygons

Let no one ignorant of geometry enter here.
—INSCRIPTION ABOVE PLATO'S ACADEMY

In this lesson you will learn what it means for polygons to be similar and about the special postulates that relate to similar triangles. You will also apply what you learned about ratios and proportions to solving problems involving similar polygons.

WHAT ARE SIMILAR POLYGONS?

When two objects have a similar appearance, they look alike, but are not exactly the same. For example, consider two laptop computers made by the same company, but with different screen sizes. A 13" laptop might have the same keyboard and overall appearance as a 17" laptop by the same manufacturer, but the 13" will be a miniature version of the 17" model. The same might go for two pairs of jeans made by the same designer, but offered in different sizes. Even though the jeans are the same style, and have the same relative appearance, one pair is just a smaller version of the other. This concept in math is referred to as "similarity." When polygons are the same overall shape, but are different sizes, they are said to be **similar**. The next figure shows two similar triangles.

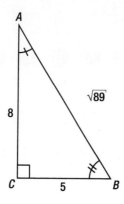
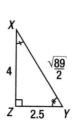

In this figure, $\triangle XYZ$ has three angles that are congruent to the angles of $\triangle ABC$. When those pairs of congruent angles are aligned in the same manner, you can see that each side of $\triangle XYZ$ is half the size of each of its corresponding side in $\triangle ABC$. *When the ratios of the corresponding sides of polygons are equal, the polygons are similar.* The symbol used to represent similar sides or similar polygons is "~". In the case of this figure, we know that $\triangle ABC \sim \triangle XYZ$, since the ratios of all the corresponding pairs of sides are equal to $\frac{1}{2}$:

$$\frac{\overline{XY}}{\overline{AB}} = \frac{\frac{\sqrt{89}}{2}}{\sqrt{89}} = \frac{1}{2}$$

$$\frac{\overline{XZ}}{\overline{AC}} = \frac{4}{8} = \frac{1}{2}$$

$$\frac{\overline{ZY}}{\overline{CB}} = \frac{2.5}{5} = \frac{1}{2}$$

....

TIP: Polygons are *similar* if their corresponding angles are congruent and the ratios of their corresponding sides are equal.

....

NAMING SIMILAR POLYGONS

When dealing with similar polygons, the order that the vertices are written in is very important. You must write the corresponding vertices of each polygon in the same order. This way, the relationship between corresponding sides is clear. Consider the two polygons in the next figure.

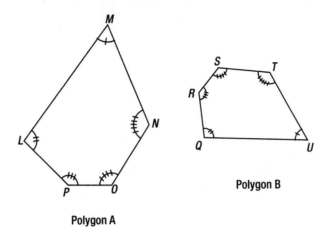

Polygon A

Polygon B

In this figure, we will name Polygon *A* by starting with ∠*M* and moving in a counterclockwise direction: *MLPON*. In Polygon *B*, ∠*U* is congruent to ∠*M*, so we will begin naming Polygon *B* with ∠*U*. If we were to again move in a counterclockwise direction, the name of Polygon *B* would be *UTSRQ*. This would imply that ∠*T* is congruent to ∠*L*, since they both are the second vertex in the name. Looking at the hash marks of the angles, you can see that ∠*L* is not congruent to ∠*T*, but instead ∠*L* ≅ ∠*Q*. Therefore, to name Polygon *B* correctly, we must begin with vertex *U* and move in a *clockwise* fashion. When written that *UQRST* ~ *MLPON*, the congruent angles match up with one another. Now, the corresponding sides of these two polygons exist between the pairs of congruent angles. Therefore, we can conclude that $\overline{RS}$ is corresponding to $\overline{PO}$.

PRACTICE 1

Use the preceding figure to answer questions **1** and **2**.

1. ∠*O* corresponds to what angle in Polygon *B*?

2. $\overline{LP}$ corresponds to what side in Polygon *B*?

3. True or false: The ratio of two pairs of sides of a polygon is sometimes equal to the ratio of the two corresponding pairs of a similar polygon.

4. True or false: When naming similar polygons, it is necessary to name both shapes by listing their vertices in the same direction—either clockwise or counterclockwise.

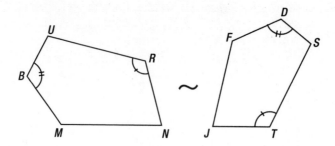

Use the similar polygons in this figure to answer questions 5–10.

5. Beginning with ∠M, name the two polygons correctly to represent their similarity.

6. ∠S corresponds to what angle?

7. ∠N corresponds to what angle?

8. What is the corresponding side to $\overline{UR}$?

9. What is the corresponding side to $\overline{FJ}$?

10. Complete the following proportion: $\dfrac{\overline{BM}}{\overline{DF}} = \dfrac{?}{\overline{JT}}$

USING PROPORTIONS WITH SIMILAR POLYGONS

Now that you know how to name similar polygons and identify their corresponding sides, you are ready to learn how to use proportions to solve for missing information in similar polygons. Consider the two triangles in the following figure.

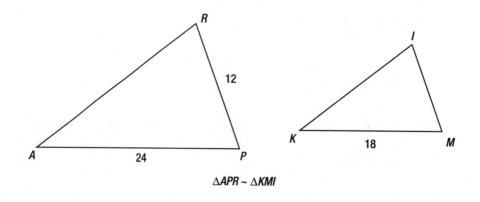

△APR ~ △KMI

The following proportion can be written about the corresponding sides of similar triangles APR and KMI: $\frac{\overline{AP}}{\overline{KM}} = \frac{\overline{RP}}{\overline{IM}}$. Then filling in the corresponding side lengths, the following proportion will be true: $\frac{24}{18} = \frac{12}{\overline{IM}}$. In the last lesson we learned to use cross multiplication to solve a proportion:

$24(\overline{IM}) = 18(12)$

$24(\overline{IM}) = 216$

$(\overline{IM}) = \frac{216}{24}$

$\overline{IM} = 9$

This is the method used to solve for missing information in similar polygons.

You probably remember that if the cross products from a proportion are not equal, then the proportion is false. This fact can be applied to similar polygons. If the proportion of corresponding sides of a polygon is false, then the polygons are not similar. For example, in the next figure, $\angle CAT$ is not similar to $\angle DOG$ since the cross products of the proportion are not equal.

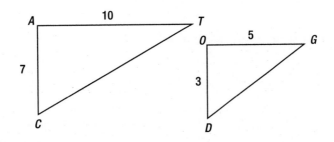

$\frac{\overline{AT}}{\overline{OG}} \overset{?}{=} \frac{\overline{AC}}{\overline{OD}}$

$\frac{10}{5} \overset{?}{=} \frac{7}{3}$

$10(3) \overset{?}{=} 5(7)$

$30 \neq 35$

PRACTICE 2

(1)

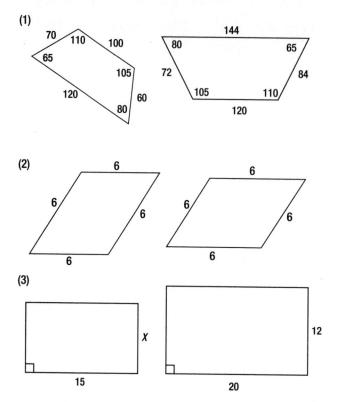

(2)

(3)

Use this figure to answer questions **1–3**.

1. Explain why the quadrilaterals in part 1 are or are not similar.

2. Explain why the quadrilaterals in part 2 are or are not similar.

3. What would the length of side x be in order to make the rectangles similar?

4. Polygon *WERT* is similar to polygon *VBNM*; *ER* = 18, *BN* = 25.2, and *WT* = 26. What is the length of side *VM*?

SIMILARITY OF TRIANGLES

There are three tests (postulates) that can be used to prove that two triangles are similar.

Test 1. Angle-Angle

If two triangles have two congruent interior angles, then they are similar. (Since the interior angles of a triangle sum to 180 degrees, as long as two angles are congruent, it can be identified that the third angles in each triangle are also congruent.)

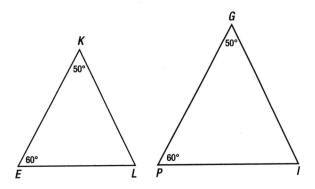

Test 2. Side-Angle-Side

If two pairs of corresponding sides of triangles are proportional, and their included angles are congruent, then the triangles are similar.

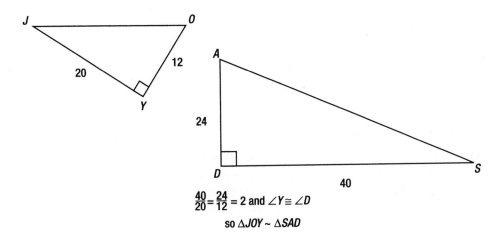

$\frac{40}{20} = \frac{24}{12} = 2$ and $\angle Y \cong \angle D$

so $\triangle JOY \sim \triangle SAD$

Test 3. Side-Side-Side

If the lengths of all pairs of corresponding sides of two triangles are all equally proportional, then the triangles are similar.

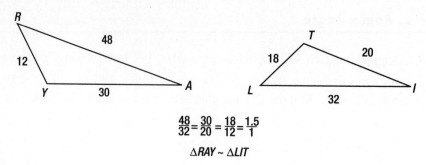

$$\frac{48}{32} = \frac{30}{20} = \frac{18}{12} = \frac{1.5}{1}$$

$$\triangle RAY \sim \triangle LIT$$

PRACTICE 3

State whether the triangles are similar and which postulate supports your answer.

1.

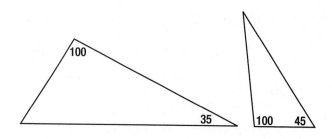

2.

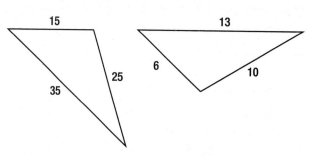

Which postulate, if any, could you use to prove that the triangles are similar?

3.

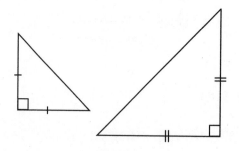

4.

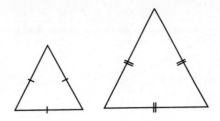

5.

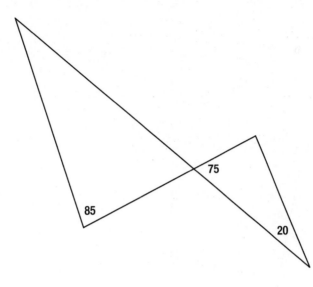

6.

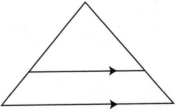

7.

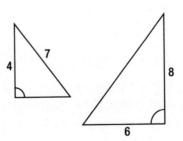

8.

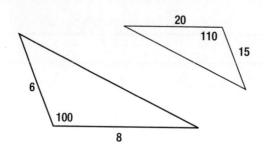

9.

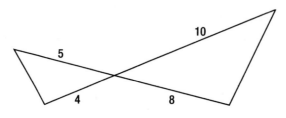

10.

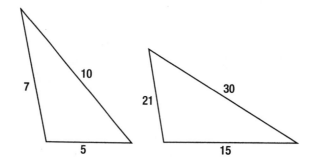

ANSWERS

Practice 1

1. $\angle S$
2. $\overline{QR}$
3. False. The ratio of two pairs of sides of a polygon is *always* equal to the ratio of the two corresponding pairs of a similar polygon.
4. False. The direction is not important—what matters is that the congruent angles from each polygon are listed in the same order.
5. *MBURN ~ FDSTJ* or *MNRUB ~ FJTSD*
6. $\angle U$
7. $\angle J$
8. $\overline{ST}$
9. $\overline{MN}$
10. $\dfrac{BM}{DF} = \dfrac{NR}{JT}$ (It would not be correct to use $\overline{RN}$.)

Practice 2

1. $\frac{72}{60} = \frac{84}{70} = \frac{120}{100} = \frac{144}{120} = \frac{1.2}{1}$ These quadrilaterals are similar because the ratios of corresponding sides of the larger quadrilateral to the smaller quadrilateral are all 1:2.

2. These polygons are similar because their corresponding angles are congruent.

3. $\frac{15}{20} = \frac{x}{12}$

   $20(x) = 15(12)$

   $20x = 180$

   $x = \frac{180}{20}$

   $x = 9$

4. $\dfrac{\overline{ER}}{\overline{BN}} = \dfrac{\overline{WT}}{\overline{VM}}$

   $\dfrac{18}{25.2} = \dfrac{26}{\overline{VM}}$

   $25.2(26) = 18(\overline{VM})$

   $655.2 = 18(\overline{VM})$

   $\frac{655.2}{18} = (\overline{VM})$

   $\overline{VM} = 36.4$

Practice 3

1. Yes. Angle-Angle. (Solve for the missing angle in each triangle, and you will see that all three pairs of angles are congruent.)

2. No. $\frac{35}{13} \neq \frac{15}{6}$

   $210 \neq 195$

3. Yes. Side-Angle-Side

4. Yes. Angle-Angle; Side-Angle-Side; or Side-Side-Side

5. Yes. Angle-Angle

6. Yes. Angle-Angle

7. No

8. No

9. Yes. Side-Angle-Side

10. Yes. Side-Side-Side

perimeter of polygons

Arithmetic is being able to count up to twenty without taking off your shoes.

—MICKEY MOUSE

In this lesson you will learn about the perimeter of polygons.

WHAT IS PERIMETER?

You will remember from Lesson 10 that the **perimeter** is the total distance around the outer edge of a shape. Landscaping professionals, carpenters, and even chefs use the concept of perimeter on a daily basis. In each of these professions, the distance around the outer edge of an object might be needed in order to purchase the correct amount of supplies for a particular project. A trick that some gardeners use to help protect their vegetable plants from harmful insects is to plant marigold flowers around the perimeter of the garden. (Insects do not like the smell and taste of marigolds, so it discourages them from entering the garden.) To find the perimeter of an irregular shape, simply add up the lengths of all the sides.

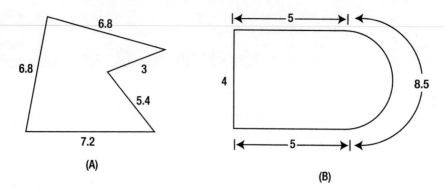

In the preceding figure, the perimeter of shape (A) would be 6.8 + 6.8 + 3 + 5.4 + 7.2 = 29.2. The perimeter of shape (B) would be 5 + 4 + 5 + 8.5 = 22.5.

PRACTICE 1

Find the perimeter of the following shapes:

1.

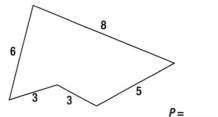

P = _____

2.

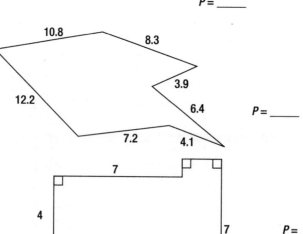

P = _____

3.

P = _____

4.

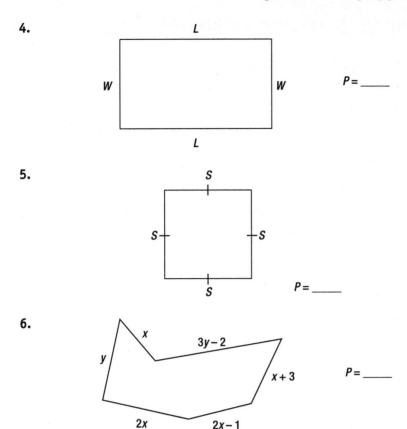

5.

P = _____

6.

P = _____

PERIMETER OF SPECIAL POLYGONS

Certain polygons have formulas for calculating the perimeter that can save you time. In the previous practice questions, you found the perimeter for two special polygons: a square and a rectangle. Another type of polygon that has a formula for perimeter is a regular polygon. Remember that when a polygon is *regular*, all of its sides are equal in length. The next figure shows the perimeter formulas you should memorize:

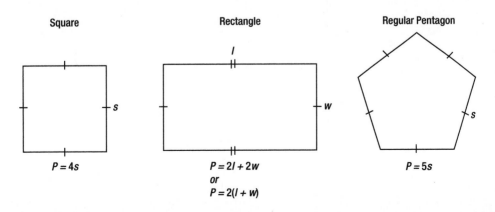

USING FORMULAS TO CALCULATE PERIMETER

Example: If a square table has one side length that is 1.5 meters long, what is its perimeter?

Solution: Since the formula for the perimeter of a square is $P = 4s$, the perimeter of this table is $4(1.5) = 6$ meters.

Example: If a rectangular rug has dimensions that are 6 feet by 8 feet, what is the total distance around the outer edge of the rug?

Solution: The formula for the perimeter of a rectangle is $P = 2l + 2w$, so the distance around the outer edge is $2(8) + 2(6) = 28$ feet.

Example: A hot tub that is a regular pentagon has a side length of 3.5 feet. What is the perimeter of the hot tub?

Solution: Since a pentagon has five sides, the formula for the perimeter of this hot tub is $P = 5s$. Therefore, $P = 5(3.5) = 17.5$ feet.

PRACTICE 2

1.

5.6

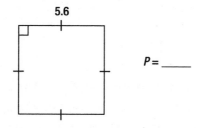

$P =$ _____

2.

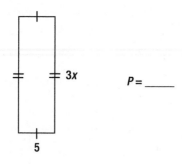

3x

5

$P =$ _____

3.

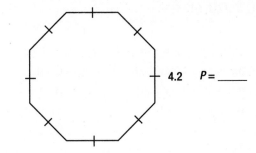

4.2 $P = $ _____

4.

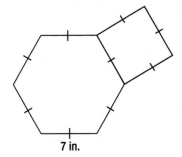

7 in.

$P = $ _____

*Find the outer edge
of this figure

5. Regular polygon $ABCDE$ if $\overline{AE} = 4$ cm.

6. Regular heptagon with side length $5xy$.

USING PERIMETER TO DETERMINE SIDE LENGTH

Now that you have algebraic formulas to use with perimeter, you can solve for side lengths of regular polygons when you are given the perimeter. For example, if you know that the perimeter of a rectangular volleyball court is 54 meters, and that the length of the court is 18 meters, you could calculate the width of the court as follows:

$P = 2l + 2w$

$54 = 2(18) + 2w$

$54 = 36 + 2w$

$18 = 2w$

$w = 9$ meters

PERIMETER WORD PROBLEMS

Word problems are common when you are dealing with perimeter. For example, consider a landscaping professional who wants to put two marigold plants along every foot of the perimeter of a square garden. If the length of the garden is 14 feet, how many marigold plants must she buy? Since the length and width of a square are equal and the formula for the perimeter of a square is $P = 4s$, the perimeter of this garden will be 4(14) =56 feet. Since the gardener wants two plants for every square foot, she will need to buy 56(2) = 112 marigolds.

PRACTICE 3

Find the length of the indicated side(s).

1.

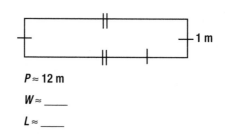

$P \approx 12$ m

$W \approx$ _____

$L \approx$ _____

2.

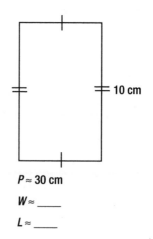

$P \approx 30$ cm

$W \approx$ _____

$L \approx$ _____

3.

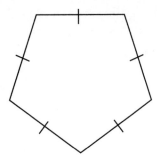

$P \approx 35$ in.

$S \approx$ _____

4.

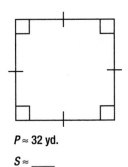

$P \approx 32$ yd.

$S \approx$ _____

5. equilateral triangle: $P = 60$ cm $S =$ _____

6. regular octagon: $P = 80$ cm $S =$ _____

7. regular hexagon: $P = 48$ cm $S =$ _____

Find answers to these practical problems.

8. Suppose you want to frame an 8-inch by 10-inch picture. How much molding will you need to buy?

9. A fence must be placed around your vegetable garden. The dimensions of the garden are 30 feet by 10 feet. How much fencing will you need?

10. You have a square dining room that you would like to trim with crown molding. If the length of the room is 17 feet, how much crown molding should you purchase?

11. Suppose you want to hang Christmas icicle lights around your house. If your house is 32 feet by 40 feet and one package of lights is 9 feet long, how many packages of lights shoud you buy?

ANSWERS

Practice 1

1. $P = 25; 3 + 3 + 5 + 8 + 6 = 25$
2. $P = 52.9; 10.8 + 8.3 + 3.9 + 6.4 + 4.1 + 7.2 + 12.2 = 52.9$
3. $P = 32$. Since all the angles in this polygon are right angles, you can identify that the total lengths of the top and bottom horizontal sides are each 9 and the lengths of the left and right sides are both 7. Therefore, there are two sides that are 7 units long and two sides that are 9 units long.
4. $P = 2L + 2W; L + W + L + W = 2L + 2W$
5. $P = 4s; s + s + s + s = 4s$
6. $P = 6x + 4y; y + x + (3y - 2) + (x + 3) + (2x - 1) + 2x = 6x + 4y$

Practice 2

1. $P = 22.4; P = 4s = 4(5.6) = 22.4$
2. $P = 10 + 6x; P = 2l + 2w = 2(5) + 2(3x) = 10 + 6x$
3. $P = 33.6; P = 8s = 8(4.2) = 33.6$
4. $P = 56$ inches; $P = 6s + 4s - 2s = 8s$ ($8s = 8(7) = 56$ inches). Notice that the one side that is in common *and* is not part of the perimeter must be subtracted twice.
5. $P = 20; P = 5s = 5(4) = 20$
6. $P = 35xy; P = 7s = 7(5xy) = 35xy$

Practice 3

1. $w = 1$ m, $l = 5$ m
2. $w = 5$ cm, $l = 10$ cm
3. $s = 7$ in.
4. $s = 8$ yd.
5. $s = 20$ cm

6. $s = 10$ cm

7. $s = 8$ cm

8. 36 in.

9. 80 ft.

10. 68 ft.

11. 16 packages

area of polygons

A man whose mind has gone astray should study mathematics.
—FRANCIS BACON

In this lesson you will review the concept of area and will learn the area formulas for squares, rectangles, triangles, parallelograms, and trapezoids.

HOW IS AREA DIFFERENT FROM PERIMETER?

The **perimeter** is the distance around a shape, which can be represented as a straight line. For example, consider a measuring tape that is wrapped around a table. It can then be pulled to form a straight line. You probably remember from Lesson 10 that the **area** of a shape is the amount of space *inside* an object, which is measured in square units. For example, let's consider the front lawn of a home as represented by the next figure.

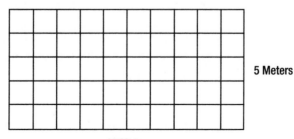

5 Meters

10 Meters

The length of the lawn is 10 meters and the width of the lawn is 5 meters. The perimeter would therefore be $2(10) + 2(5) = 30$ meters. The perimeter of 30 meters might be used to help a homeowner figure out how much fencing to purchase to make a closed yard for his dog. The *area* of the lawn is the amount of space *inside* the fenced-off area and is measured in square meters. For every meter of length, there are five meters of depth to the lawn. This means that 10 *times* 5, or 50, square meters will fill the space of the lawn. This measurement would be helpful if the homeowner was purchasing square meters of sod to plant in the front lawn. When representing area, use the squared exponent to show that area is a two-dimensional measurement. In this case, it would be incorrect to write that the area is 50 meters, since that implies that the area is a straight line. It would be *correct* to write that the area is equal to 50 m^2 or 50 meters^2.

AREA FORMULAS FOR RECTANGLES, SQUARES, AND TRIANGLES

We already saw in the preceding example that the area of the rectangular lawn was calculated by multiplying the 10 meters of length by the 5 meters of width. The general formula for the area of a rectangle is length *times* width.

TIP: The *area of a rectangle* = (length) × (width), or $A = lw$

Since the length and width in squares are the same size, the formula for the area of a square is the length of one of its sides, times itself. In squares it is typical to refer to the side length as *s* for side, instead of using *length* or *width* as is used in rectangles.

TIP: The *area of a square* = (side) × (side), or $A = s^2$

The formula for the area of a triangle is a concept that was presented in Lesson 10. Now that you know the formula for the area of a rectangle, it is helpful to see how that relates to the area of a triangle.

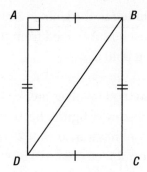

Looking at this figure, you can see that when a diagonal is drawn in rectangle *ABCD*, two congruent triangles are formed. (We know they are congruent by the Side-Side-Side postulate.) Since the area of the rectangle is length *times* width, you can identify that the area of one of the triangles would be half of length *times* width. Replacing "base" for the "width," and "height" for the "length," this brings us to the formula for area of a triangle presented in Lesson 10: Area $= \frac{1}{2} \times$ (base) $\times$ (height).

TIP: The *area of a triangle* $= \frac{1}{2} \times$ (base) $\times$ (height) or $A = \frac{1}{2}bh$

AREA OF PIECEWISE FIGURES

Sometimes, a figure must be broken up into separate polygons in order to identify its area. Consider the next figure, which represents the lawn around a square patio. In order to find the area of the lawn there are two methods.

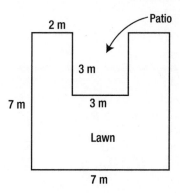

Method 1: Find the area of the entire figure and then subtract the smaller portion that is not included. Here, the larger shape is a square that is 7 meters long by 7 meters wide. The area of that square is therefore 49 m². The smaller

patio section that is not included in the lawn is 3 meters by 3 meters. The area of the patio is 9 m². This must be subtracted from the larger area: 49 m² – 9 m² = 40 m². The area of the lawn is 40 m².

    Method 2: The second method involves breaking the figure up into separate polygons as illustrated in this next figure. Sections *A* and *C* will both have areas of 6 m² and section *B* will have an area of 28 m². Adding these three separate sections together, the combined area will be 40 m².

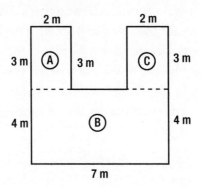

PRACTICE 1

Find the area of each polygon:

1.

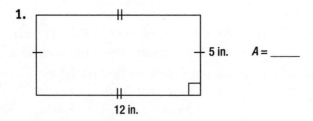

$A =$ _____

2.

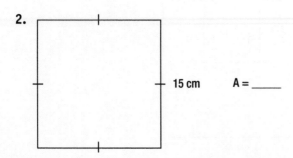

$A =$ _____

3.

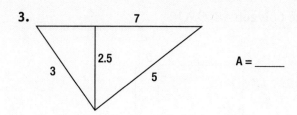

A = _____

4.

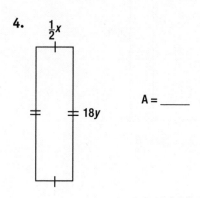

A = _____

5.

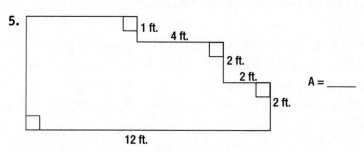

A = _____

6. Find the area of the polygon *BCDEFGHI*.

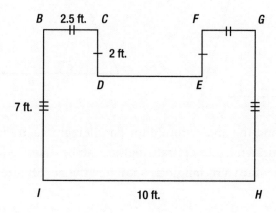

7. Find the area of the concave polygon *MNOPQR*.

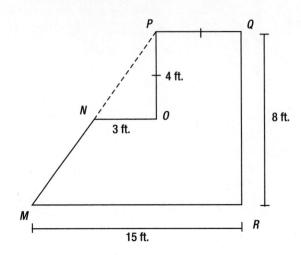

AREA OF PARALLELOGRAMS AND TRAPEZOIDS

The relationship between the base and the height of a parallelogram is the same that is found in triangles. Any side of a parallelogram can be called the base. The corresponding altitude must be drawn from that base to the opposite vertex so that it forms a right angle. This altitude is called the height and is illustrated in the next figure.

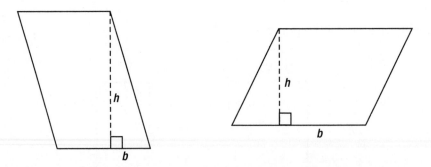

To understand the area formula for parallelograms, notice that once the height has been drawn in, any parallelogram can be dissected into two pieces and rearranged to form a rectangle as shown in the next figure.

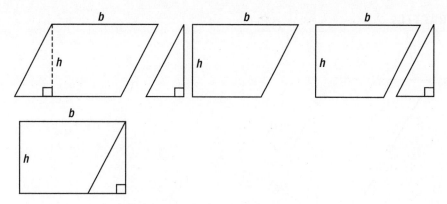

Looking at this figure, it is probably clear to you that the area of a parallelogram can be determined by multiplying the base times the height.

TIP: The *area of a parallelogram* = (base) × (height) or $A = bh$

In a trapezoid, the bases are parallel, but not congruent, as in a parallelogram. In order to find the area of a trapezoid, the average of the two bases is multiplied by the height. (You might remember from Lesson 16, that the average of a trapezoid's bases is called its *median*.) This is more commonly written as one-half the product of the height times the sum of the bases. This is shown in the next figure.

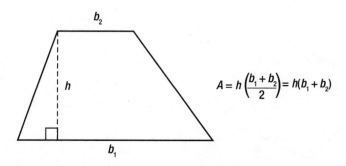

$$A = h\left(\frac{b_1 + b_2}{2}\right) = h(b_1 + b_2)$$

TIP: The *area of a trapezoid* = $\frac{1}{2}$(height)(base$_1$ + base$_2$), or $A = \frac{1}{2}(h)(b_1 + b_2)$

PRACTICE 2

1.

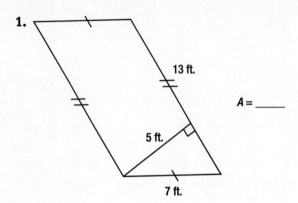

13 ft.

5 ft.

7 ft.

A = _____

2.

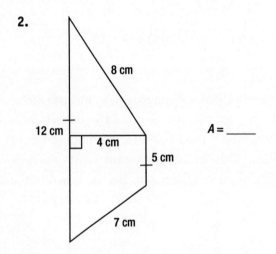

8 cm

12 cm

4 cm

5 cm

7 cm

A = _____

3.

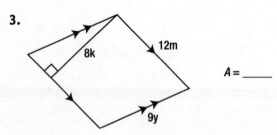

8k

12m

9y

A = _____

4.

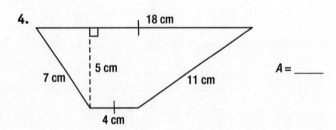

18 cm

7 cm

5 cm

11 cm

4 cm

A = _____

Find the area of each figure below.

5. Find the area of quadrilateral *ABCD*.

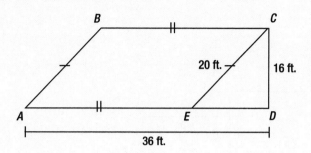

6. Find the area of polygon *RSTUV*.

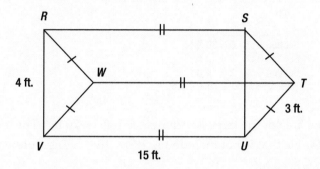

ANSWERS

Practice 1

1. 60 in.^{2}
2. 225 cm^2
3. 8.75 units^2
4. $9xy$ units^2
5.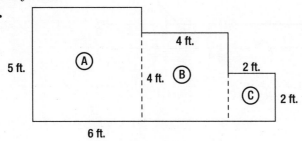

Section A : $5 \cdot 6 = 30$ ft.^{2}
Section B : $4 \cdot 4 = 16$ ft.^{2}
Section C : $2 \cdot 2 = \ \ 4$ ft.^{2}
Total Area $= 30 + 16 + 4 = 50$ ft.^{2}

6. 60 ft.^{2}. The area of the largest rectangle is $10(7) = 70$ ft.^{2}. The area of the inset rectangle that is not included is $2(5) = 10$ ft.^{2}. The difference of the two shapes is 60 ft.^{2}.

7.

Section A:
$$\frac{1}{2}(11)(8) - \frac{1}{2}(3)(4)$$
$$44 - 6 = 38 \text{ ft.}^2$$

Section B:
$8(4) = 32$ ft.^{2}

Total $= 38$ ft.^{2} $+ 32$ ft.^{2} $= 70$ ft.^{2}

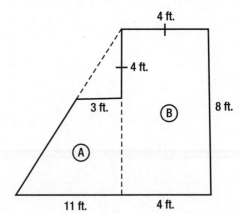

Practice 2

1. $A = 5(13) = 65$ ft.^{2}

2. $A = \frac{1}{2}(4)(12 + 5) = 34$ cm^2

3. $A = 8(12) = 96$ m^2

4. $A = \frac{1}{2}(5)(18 + 4) = 55$ cm^2

5. $(\overline{ED})^2 = 20^2 - 16^2$

$(\overline{ED})^2 = 144$

$(\overline{ED}) = 12$

$\overline{AE} = 36$ ft. $- 12$ ft. $= 24$ ft.

$A(AECB) = 24(16) = 384$ ft.^{2}

$A(ABCD) = 384 + 96 = 490$ ft.^{2}

6.

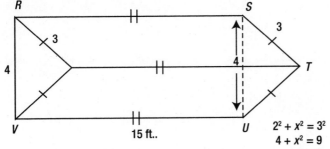

$2^2 + x^2 = 3^2$
$4 + x^2 = 9$
$x^2 = 5 \quad x = \sqrt{5}$

A (Rectangle *RSUV*) = (15 ft.)(4 ft.) = **60 ft.^{2}**

Next calculate area of triangle:

$A = \frac{1}{2}(4)(\sqrt{5}) = 2\sqrt{5}$

Total Area = 60 ft.^{2} + 2$\sqrt{5}$ ft.^{2}

introduction to circles

Everything tries to be round.
—BLACK ELK

In this lesson you will study concepts relating to circles, such as radii, diameter, arcs, secants, central angles, and inscribed angles.

BASIC TERMS OF CIRCLES: RADII, DIAMETER, AND CHORDS

A **circle** is a collection of points that are an equal distance from one singular point. That one point is the **center** of a circle and it is what defines the name of any given circle. If the center point of a circle is labeled *F*, it will be referred to as *circle F*.

....

TIP: Circles are named by their center point.

....

The distance from the center of the circle to any point on the edge of the circle is called the **radius** of the circle. A circle has an infinite number of radii, which are all the same length ("**radii**" is the plural of "radius"). In the next figure, $\overline{FA}$, $\overline{FB}$, and $\overline{FC}$ are three congruent radii in circle *F*.

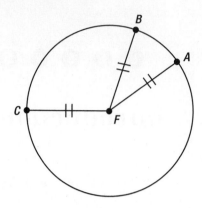

··

TIP: The distance from the center of the circle to a point on the circle is called the *radius*.

··

Another important term when you are dealing with circles is **chord**. A chord is a line segment that extends from one point on the circle to another point on the circle. In the next figure, line segments $\overline{PQ}$ and $\overline{VW}$ are both chords in circle G.

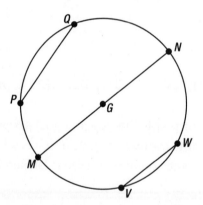

··

TIP: A *chord* is a line segment that extends from one point on the circle to another point on the circle.

··

Can you see that $\overline{MN}$ in the preceding figure is also a chord? When a chord passes through the center of a circle, it is called a **diameter**. Since the diameter is essentially made up of two radii extending from the center in opposite directions, the diameter will always be *twice the length* of a radius.

TIP: A *diameter* is a special chord that passes through the center of a circle. The length of the diameter is twice the radius.

ARCS IN CIRCLES

When you hear the word *arc* you should think of something curved. In circles, an **arc** is a curved section of the outside of the circle that is defined by two points on the circle. Sometimes these points will be the endpoints of a chord. Arcs are written similarly to chords, except the line over them is curved. For example, in this figure, chord $\overline{PQ}$ defines arc PQ, which is written $\overset{\frown}{PQ}$. Similarly, VW creates $\overset{\frown}{VW}$.

TIP: A *major arc* is the longer curved distance around a circle formed by two points on the circle. A *minor arc* is the shorter curved distance formed by two points on the circle.

When you're reading $\overset{\frown}{VW}$, you cannot be certain if what is being referred to is the smaller counterclockwise distance from V to W or the longer clockwise distance from V to W. To clarify this, the term **minor arc** is used to refer to the shorter distance of an arc. **Major arc** is used to indicate the longer arc created by two points. Can you tell that as the diameter, $\overline{MN}$ is defining a special type of arc? $\overline{MN}$ divides the circle into two arcs of equal distance: $\overset{\frown}{MN}$ and $\overset{\frown}{NM}$. Each of these arcs is called a **semicircle**.

TIP: A *diameter* forms an *arc* that is a *semicircle*.

Sometimes, in order to avoid confusion when you're naming arcs, you use a third point on the circle to show the direction of the arc. This three-letter naming is most commonly used for major arcs. For example, in the following figure $\overset{\frown}{AB}$ refers to the clockwise minor arc created by chord $\overline{AB}$. $\overset{\frown}{AYB}$ refers to the counterclockwise major arc created by chord $\overline{AB}$.

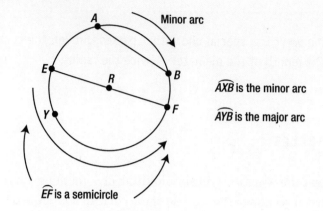

$\overset{\frown}{AXB}$ is the minor arc

$\overset{\frown}{AYB}$ is the major arc

$\overset{\frown}{EF}$ is a semicircle

PRACTICE 1

Use this figure to answer the following questions.

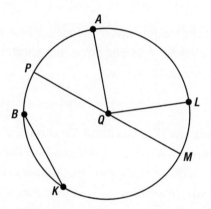

1. What is the name of the circle in this figure?

2. Line segment $\overline{QL}$ is called the _____ of the circle.

3. $\overline{QL} \cong$ _____.

4. The shorter clockwise distance along the circle from point L to point M is called the _____.

5. The longer counterclockwise distance along the circle from point L to point M is called the _____.

6. $\overline{PM}$ is a _____.

7. What is true about the length of $\overline{PM}$?

8. $\overline{BK}$ is called a _____.

9. What is one way to name the minor arc formed by points A and M?

10. What is one way to name the major arc formed by points A and M?

11. $\overset{\frown}{PM}$ forms what kind of arc?

DEGREES IN CIRCLES

The length of an arc can be represented in units such as centimeters or inches, but more commonly, the length is represented as a measure of degrees. A complete revolution around a circle contains 360°. A semicircle contains 180° since it is half the distance around a circle. A *minor arc* is any arc that measures less than 180° and a *major arc* is any arc that measures more than 180°.

....

TIP: A circle contains 360°. A semicircle contains 180°.

....

CENTRAL ANGLES

A **central angle** is an angle whose vertex is the center of the circle with two endpoints on the circle. Central angles are formed by two radii of a circle. $\angle AQL$ in the figure in Practice 1 is a central angle of circle Q.

....

TIP: A *central angle* is an angle formed by two radii. Its vertex is the center of the circle and whose endpoints sit on the circle.

....

The minor arc formed by a central angle is referred to as the ***intercepted arc***. There is a special relationship between the measure of a central angle and its intercepted arc. To explore this relationship, let us first look at the angle created by the diameter AB in the next figure.

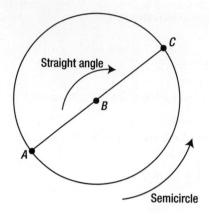

You know that a straight line contains 180°. You also know from the discussion above, that a semicircle also contains 180°. In this case, $\overset{\frown}{AC}$ created by central angle *ABC* has the same measure as ∠*ABC*.

Take a look at the central angle in the next figure, which has been drawn in as a 90° right angle.

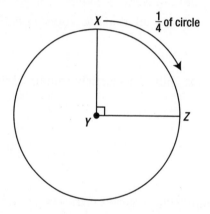

Can you tell that minor arc $\overset{\frown}{XZ}$ defines $\frac{1}{4}$ of the outside of the circle? Since a circle contains 360°, $\frac{1}{4}$ of a circle contains 90°. Therefore, the measure of intercepted arc *XZ* equals 90°. In these two examples the measure of the central angle equals the measure of its intercepted arc. This rule holds true for all central angles.

....

TIP: The measure of a central angle is equal to the measure of its intercepted arc.

....

INSCRIBED ANGLES

When the vertex of an angle is on the circle and the angle's sides are two chords within the circle, that angle is called an **inscribed angle**. The measure of an inscribed angle is one-half the measure of its intercepted arc. For example, in the next figure, $m\angle ABC = 45°$ and $\overarc{AC} = 90°$. In the same figure, if $m\angle XYZ = 20°$, then $\overarc{XZ} = 40°$.

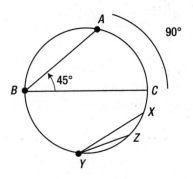

....

TIP: An *inscribed angle* has its vertex and endpoints on the circle. Its measure is half that of the arc it creates.

....

You now have the basics needed to solve a variety of problems relating to arcs and angles within circles. In summary, here are the key points that you have learned in this lesson:

- A circle contains 360°.
- A diameter creates two semicircles that each contain 180°.
- The measure of a central angle equals the measure of its intercepted arc.
- An inscribed angle has a measure that is one-half of the measure of the arc it creates.

Use this figure while you're looking through the examples that follow.

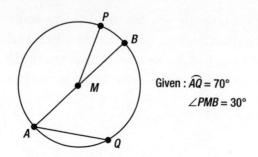

Given : $\overset{\frown}{AQ} = 70°$
$\angle PMB = 30°$

Example 1: Find the measure of intercepted arc $\overset{\frown}{BQ}$.

Solution: Since $\overline{AB}$ is a diameter, we know that $\overset{\frown}{AB}$ is a semicircle measuring 180°. Since $\overset{\frown}{AQ}$ measures 70°, $\overset{\frown}{BQ}$ must measure 110°.

Example 2: Find the measure of $\angle BAQ$.

Solution: Since $\overset{\frown}{BQ}$ measures 110°, you know that inscribed angle $\angle BAQ$ measures half of that, so $\angle BAQ = 55°$.

Example 3: Find the measure of intercepted arc $\overset{\frown}{PB}$.

Solution: Since central $\angle PMB = 30°$, its intercept arc will also have the same measure of 30°.

PRACTICE 2

Use this figure to answer questions 1–4.

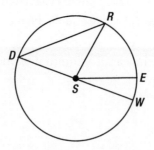

1. If $m\angle RSE = 60°$, what is the measure of $\overset{\frown}{RE}$?

2. If the measure of $\overset{\frown}{WE} = 20°$, what is the $m\angle RDW$?

3. Using the information in questions **1** and **2**, what is the measure of $\overgroup{DR}$?

4. Disregard the information given in questions **1** through **3**. Using the figure above, if $m\overgroup{DR} = 6x$, $m\overgroup{RE} = 4x$, $m\overgroup{EW} = 2x$, and $m\overgroup{DW} = 12x$, find the measure of each of the arcs.

5.

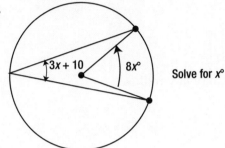

Solve for $x°$

6.

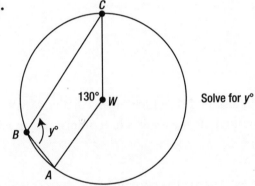

Solve for $y°$

7.

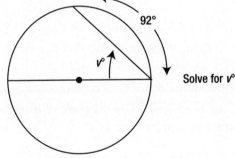

Solve for $v°$

ANSWERS

Practice 1

   1. Circle Q
   2. radius
   3. $\overline{QL} \cong \overline{QA} \cong \overline{QP} \cong \overline{QM}$
   4. minor arc $\overset{\frown}{LM}$
   5. major arc $\overset{\frown}{LM}$
   6. diameter
   7. The diameter $\overline{PM}$ is twice the length of any radius.
   8. chord
   9. $\overset{\frown}{AM}$
   10. $\overset{\frown}{APM}$, $\overset{\frown}{ABM}$, or $\overset{\frown}{AKM}$
   11. $\overset{\frown}{PM}$ forms a semicircle.

Practice 2

   1. $\overset{\frown}{RE} = 60°$
   2. $m\angle RDW = 40°$, $m\angle ESW$ would also be 20°, which means that $m\angle RSW = \overset{\frown}{RW} = 80°$. Since $\angle RDW$ is inscribed, its measure is half of its intercepted arc, so it is 40°.
   3. $\overset{\frown}{DR} = 180° - 80° = 100°$.
   4. Since the sum of all the arcs is 360°, write the equation $6x + 4x + 2x + 12x = 360°$. Solving for x gives $x = 15°$, so the measure of the arcs is 90°, 60°, 30°, and 180°.
   5. Since the inscribed angle measures $3x + 10$, we know its intercepted arc will be twice that, or $6x + 20$. The central angle $8x$ shows that the same intercepted arc will measure $8x$. Therefore, $6x + 20 = 8x$ and $x = 10$.
   6. Central $\angle CWA$ shows that inscribed arc $\overset{\frown}{CBA} = 130°$. Therefore, major arc $\overset{\frown}{CA}$ will measure 230° (since $360° - 130° = 230°$.) The measure of inscribed angle $\angle CBA$ will be half of its intercepted arc, so $y = 115°$.

LESSON 22

pi and the circumference of circles

A circle is a round straight line with a hole in the middle.
—MARK TWAIN

In this lesson you will learn about the irrational number, π, and how it connects with the circumference of circles.

WHAT IS π?

The irrational number π (which is spelled *pi* and pronounced "pie"), is a key concept when working with circles. It relates to the **circumference** of a circle, which is the perimeter around the outside of a circle.

TIP: The distance around a circle is called its *circumference*.

More than 2,000 years ago, mathematicians worked to find a shortcut to calculate a circle's circumference, by using the length of the circle's diameter. This may seem like a nearly impossible task, but you do not have to be an expert mathematician to investigate this relationship. You could do this yourself with just a ruler, a piece of string, and several different-sized cylinders, such as a toilet paper roll, a can of soup, and a salad bowl. If you measured the distance around each cylinder and divided it by the cylinder's diameter, you would find that the quotient of the $\frac{\text{circumference}}{\text{diameter}}$ is always just a little bit larger than 3. This would be a great start to estimating π, which is defined as the ratio of a circle's circumference to its diameter. This relationship is the same in all circles, and the most commonly used approximation of the ratio between a circle's circumference and its diameter is 3.14. Pi does not stop at 3.14 though; π is an irrational number that carries on for billions of digits and does not end.

TIP: Pi, or π, is an irrational number equal to the ratio $\frac{\text{circumference}}{\text{diameter}}$ of a circle. π is commonly approximated as 3.14 (or $\frac{22}{7}$).

CIRCUMFERENCE OF A CIRCLE

Now that the relationship between a circle's circumference and diameter has been determined, it is possible for you to come up with a formula used to calculate circumference:

$$\frac{\text{circumference}}{\text{diameter}} = \pi$$
$$\frac{\text{circumference} \times (\text{diameter})}{\text{diameter}} = \pi \times (\text{diameter})$$
$$\text{circumference} = \pi \times (\text{diameter})$$
$$C = \pi d$$

Since the diameter is always twice the length of the radius, r, another way to express circumference is $C = 2\pi r$.

TIP: Circumference = πd or $2\pi r$.

Since π is an irrational number, answers relating to circumference are often left in terms of "π," instead of being evaluated with π as 3.14. In general, questions state whether to leave your answer in terms of π, or to round it to the nearest tenth or hundredth. When you're rounding your answer, it is good practice to use the symbol for approximation, which is $\approx$, instead of an equals sign. Refer to the next figure for the following examples.

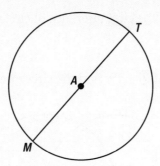

Example 1: Find the circumference of circle A, if $MT = 6$ (leave your answer in terms of π).

Solution: Since circumference $= \pi d$, $C = 6\pi$.

Example 2: Find the circumference of circle A, if $\overline{AT} = 5$ (round your answer to the nearest hundredth).

Solution: Since circumference $= 2\pi r$, $C \approx 2(3.14)(5) \approx 31.4$.

Example 3: If the circumference of circle $A = 36\pi$, find the length of $\overline{MA}$.

Solution: Since $C = 2\pi r$, $36\pi = 2\pi r$ and $\frac{36\pi}{2\pi} = r$, so $\overline{MA} = 18$.

PRACTICE 1

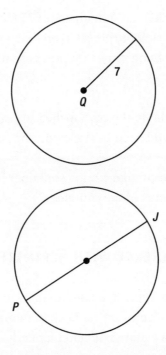

1. In the first figure, find the circumference of circle Q to the nearest hundredth.

2. In the second figure, if $\overline{PJ} = 19$, find the circumference of the given circle in terms of π.

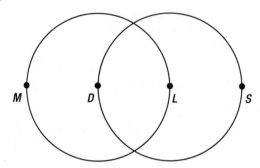

3. In this figure, $\overline{DL}$ is the radius of both circle D and circle L. If $\overline{MS} = 15$, find the circumference of circle D to the nearest hundredth.

4. Using the same figure from question 3, if $\overline{MD} = 2x$, what is the circumference of circle L in terms of π and x?

5. Using the figure from question 3, if the circumference of circle L is (14)(v)(π), what is the length of $\overline{MS}$ in terms of v?

6. Your circular rose garden has a diameter of 25 feet. Fencing is sold by the yard at Francine's Fencing Surplus. Given that a yard equals three feet, how many yards of fencing must you purchase in order to encircle your rose garden?

7. If the radius of a circular hatbox is 8 inches, approximately how many inches of ribbon will you need to circle your hatbox one time?

8. The circumference of a circular plastic container is approximately $39\frac{1}{4}$ inches. What is the radius of the container?

COMBINING CIRCUMFERENCE WITH PERIMETER

In the real world, many objects are a combination of two or more shapes put together. A corner office desk may be made up of two different-sized rectangles, or a coffee table might be a square with circular ends. It is important to be able to do calculations when you're working with these more complex shapes. Often, you will need to identify which portion of the circle you're using—using half of the circle or one-quarter of the circle are common. In addition, you will need to decide if there are parts of the polygon that should be omitted when you're calculating the perimeter. These problems are best done by breaking them up into smaller, separate parts. Consider the following problem that accompanies the next figure.

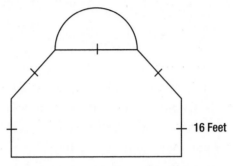

Example: The main hall of the building is the top part of a regular octagon. Its vertical walls, slanted walls, and ceiling are all congruent. The top of the building is a semicircle. If some students were trying to calculate how many feet of twinkle lights were needed to line the outside of this building, not including the floor, how would this be done?

Solution: Since there are 4 edges that are 16 feet each, the straight edges will need 4(16) = 64 feet of lights. Last, the semicircle on the top will need $\frac{1}{2}$ the circumference of lights. $C = \pi d = 16\pi$, so one-half of this is 8π. 8π estimates to just over 25 feet, so if the students bought 64 + 26 = 90 feet of lights, they would be able to line the outside edges of this building.

PRACTICE 2

1.

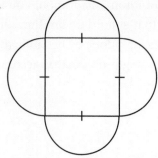

Melissa is giving a party and has designed invitations that will look like the petals of a flower when the semicircular edges are folded over the square invitation. She wants to line the outside edge of the four semicircular parts with silk ribbon. If the inner square part of the invitation measures 5 inches by 5 inches, approximately how many inches of ribbon will she need to line each invitation?

2.

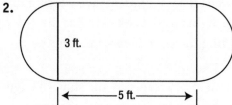

Fred is hosting a party for his two-year old daughter. He needs to buy foam to line the outside edge of a coffee table to protect the heads of any kids who might fall against the table. The table has a rectangular section that is 3 feet wide by 5 feet long, and at each edge of the table there is a semicircular section. Approximately how many feet of foam does Fred need to buy?

3.

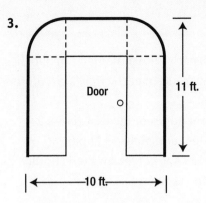

Door

11 ft.

10 ft.

A door is 4 feet wide by 8 feet tall and it is symmetrically centered in an archway that is 10 feet wide by 11 feet tall. If Juan wants to line the outside edge of the archway with tiles, how many feet of tiles will he need to complete the job? (The edge to be lined with tiles has been darkened.)

ANSWERS

Practice 1

1. $C = 2\pi r = 2\pi 7 \approx 14(3.14) \approx 43.96$
2. $C = \pi d = 19\pi$
3. Since $\overline{MS} = 15$, then $\overline{DL} = 5$ since $\overline{MS}$ is made up of three congruent radii. Therefore, the circumference of circle $D = 2\pi r \approx 10(3.14) \approx 31.4$.
4. Since $\overline{MD} = 2x$, then $\overline{DS} = 4x$ and the circumference of circle $L = 4x\pi$.
5. Since $C = 2\pi r$ then $(14)(v)(\pi)$. Divide both sides by 2π to get $r = 7v$. $\overline{MS}$ is three radii in length, so $\overline{MS} = 3(7v) = 21v$.
6. $C = \pi d = 25(3.14) = 78.5$ feet. Now change this into yards by dividing it by 3 feet: $\frac{78.5}{3}$ 26.166. Therefore, you need to purchase 27 yards of fencing to encircle the garden.
7. $C = 2\pi r$, so $C = 2\pi 8 \approx 16(3.14) \approx 50.24$ inches
8. $C = 2\pi r$, so $39.25 = 2\pi r$
   $2\pi r \approx 2(3.14)r \approx 39.25$, so $r \approx \frac{39.25}{6.28} \approx 6.25$ inches

Practice 2

1. Melissa's invitation has four semicircles, each with a diameter of 5 inches. So the distance around the outside is (4 parts)($\frac{1}{2}$ of each circle's circumference) = $(4)(\frac{1}{2})\pi d = 2\pi 5 = 10\pi \approx 31.4$ inches.

2. The foam needed to cover the two straight edges of the table is $5 \times 2 = 10$ feet. The circumference of the two outer semicircles is (2 parts)($\frac{1}{2}$ of each circle's circumference) = $(2)(\frac{1}{2})\pi d = 1\pi 3 \approx 9.42$ feet. So Fred should buy 19.5 or 20 feet of foam to surround the table.

3.

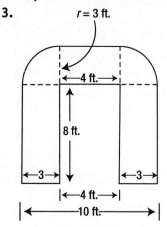

The outside of the archway can be broken up into three rectangles and two quarter-circles. These two quarter-circles can be combined to form a semicircle with a radius of 6. The circumference of this semicircle is ($\frac{1}{2}$ of the circumference) = $(\frac{1}{2})2\pi r = 1\pi 3 \approx 9.42$ feet. The vertical edges of the archway are each 8 feet long, so 16 feet of tiles will be needed there. Last, the top edge of the archway is 4 feet. The total outside border of the archway is $9.42 + 16 + 4 \approx 29.42$ feet.

the area of circles

In mathematics you don't understand things.
You just get used to them.
—JOHANN VON NEUMANN

In this lesson you will learn how to apply π to the area of circles. You will also learn how to determine the area and perimeter of sectors of circles when you're given central angle measurements.

AREA OF A CIRCLE

In a previous lesson, we showed visually that the area of a triangle is $\frac{1}{2}$(base)(height) by demonstrating how a rectangle can be broken down into two congruent triangles. We also illustrated why the area of a parallelogram is (base)(height). It is also possible to use a visual to help understand the formula for the area of a circle. Consider the circle that has been divided into gray-and-white triangles in the next figure.

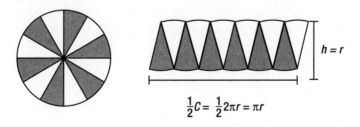

$$\tfrac{1}{2}C = \tfrac{1}{2}2\pi r = \pi r$$

TIP: The area of a circle is equal to πr^2.

It is important to notice that the formula for the area of circles uses radius and not diameter. Therefore, if you are given only the diameter of a circle, you must first divide it in half before using the area formula. (Remember that the radius *squared* is not the same thing as the radius times 2. You *cannot* use diameter in place of r^2.)

Use the circles in this figure for examples 1 and 2 that follow.

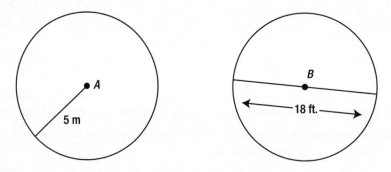

Example 1: Find the area of circle A to the nearest tenth of a meter.

Solution: Area $= \pi r^2 = \pi \, (5^2) \approx 3.14(25) \approx 78.5 \text{ m}^2$.

Example 2: Find the area of circle B. Keep your answer in terms of π.

Solution: The radius of circle B is $\frac{18}{2} = 9$ feet. Area $= \pi r^2 = \pi(9^2) = 81\pi \text{ ft.}^2$.

PRACTICE

1. In the next figure, what is the area of circle H? Keep your answer in terms of π.

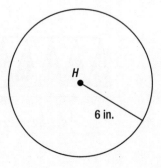

2. In the next figure, what is the area of circle *I*? Keep your answer in terms of π.

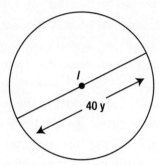

3. In the shaded figure, what is the area of the shaded half of the circle? Round your answer to the nearest tenth.

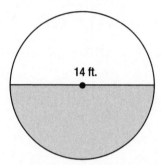

4. Victoria makes Parker Elizabeth a round birthday cake that had a diameter of 16 inches. What is the area of the top of the cake? (Round your answer to the nearest tenth.)

5. Michael is helping Maddie Winn paint a round sun on her bedroom wall. Maddie Winn wants the sun to go from the ceiling to halfway down the wall toward the floor. If her ceiling is 9 feet high, how many square feet of paint will be used to paint the sun? (Round your answer to the nearest tenth.)

6. In the previous lesson we calculated the number of feet of fencing needed for a circular rose garden with a diameter of 25 feet. How many square feet of mulch will be needed to cover the garden? (Round your answer to the nearest tenth.)

7. If the area of cover for a circular children's pool is 38.5 ft.^{2}, what is the radius of the pool?

8. The area of a circular stage is 452.2 m^2. What is the diameter of the stage?

FINDING THE AREA OF A SECTOR OF A CIRCLE

A **sector** is a part of a circle bound by two radii and their intercepted arc. A sector looks like a piece of pie, or an ice cream cone. Sometimes, you will to need calculate the area or perimeter of just a sector of a circle. Remember that a central angle has the same degree measurement as its intercepted arc. Let us consider a circle that has a central angle of θ. The sector enclosed by that angle will occupy a portion of that circle that is equal to the *measure of angle* divided by the *total number of degrees* in the circle: $\frac{\theta°}{360°} \cdot \frac{\theta°}{360°}$ represents the fractional part of the area you will be solving for. Therefore, you can always calculate the area of the sector of a circle by multiplying the *fractional part* of the circle by the *total area* of the circle: $(\frac{\theta°}{360°})(\pi r^2)$.

TIP: Area of a sector of a circle $= (\frac{\theta°}{360°})(\pi r^2)$

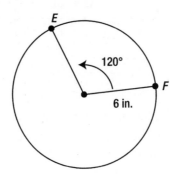

Example: Find the area of the sector in this figure.

Solution:

$A(\text{sector}) = (\frac{\theta°}{360°})(\pi r^2)$

$A(\text{sector}) = (\frac{120°}{360°})(\pi 6^2)$

$A(\text{sector}) = (\frac{1}{3})(36\pi)$

$A(\text{sector}) = 12\pi \text{ in.}^2$

FINDING THE LENGTH OF A SECTOR'S ARC

The same principle can be used to calculate the length of the intercepted arc formed by a central angle. Multiply the fractional part of the circle by the total circumference of the circle: $(\frac{\theta°}{360°})(\pi d)$.

TIP: Length of a sector's arc $= \left(\frac{\theta°}{360°}\right)(\pi d)$

Example: Find the length of the intercepted arc $\overparen{EF}$ in the preceding figure.

Solution

$C(\text{arc}) = \left(\frac{\theta}{360°}\right)(\pi d)$

$C(\text{arc}) = \left(\frac{120°}{360°}\right)(\pi 12)$

$C(\text{arc}) = \left(\frac{1}{3}\right)(12\pi)$

$C(\text{arc}) = 4\pi$ inches

PRACTICE 2

1.

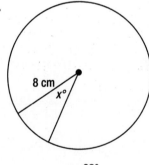

$x = 32°$

Find the area to the nearest hundredth of the sector defined by angle x.

2. Using the figure from question **1**, find the length of the intercepted arc defined by angle x. Round it to the nearest tenth.

3.

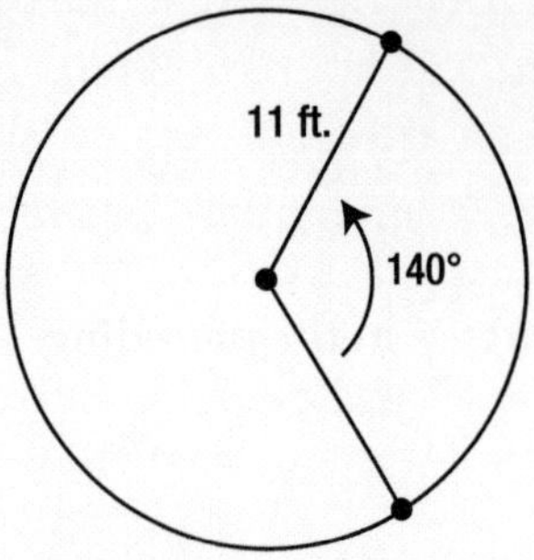

This figure shows a circular garden space. The sector created by the 140-degree angle will have woodchips and benches for people to sit on. The remainder of the garden space will have grass. How many square feet of this area will be covered in grass? Round to the nearest whole foot.

ANSWERS

Practice 1

1. Area $= \pi r^2 = \pi(6^2) = 36\pi$ in.^{2}.
2. The radius of circle B is $\frac{40}{2} = 20$ y. Area $= \pi r^2 = \pi(20^2) = 400\pi$ y^2.
3. The radius of the circle is $\frac{14}{2} = 7$ ft. Area $= \pi r^2 = \pi(7^2) = 49\pi$ ft.$^2 \approx 153.86$ ft.^{2} The area of the shaded half is therefore 76.9 ft.^{2}.
4. The radius of the cake is $\frac{16}{2} = 8$ inches. Area $= \pi r^2 = \pi(8^2) = 64\pi$ in.$^2 \approx$ 200.96 in.$^2 \approx 201.0$ in.^{2} when rounded to the nearest tenth.
5. Since the ceilings are 9 feet tall, and the sun will go halfway down the wall, it will have a diameter of 4.5 feet. The sun will therefore have a radius of 4.5 feet divided by two, which will be 2.25 ft. Area $= \pi r^2 = \pi(2.25^2) = 5.0625\pi$ ft.$^2 \approx 15.9$ ft.^{2}. Michael will need to buy enough paint to cover approximately 16 square feet.
6. The rose garden has a radius of 12.5 feet. Area $= \pi r^2 = \pi(12.5^2) \approx 490.6$ ft.^{2}
7. Area $= \pi r^2$; $38.5 = \pi r^2$; $r^2 = \frac{38.5}{3.14} = 12.26$, so $r \approx 3.5$ feet.
8. Area $= \pi r^2$; $452.2 = \pi r^2$; $r^2 = \frac{452.2}{3.14} = 144.01$, so $r \approx 12$ feet and $d \approx 24$ feet.

Practice 2

1. $A(\text{sector}) = (\frac{\theta°}{360°})(\pi 8^2)$
   $A(\text{sector}) = (\frac{32°}{360°})(64\pi)$
   $A(\text{sector}) = 17.86$ cm^2
2. $C(\text{arc}) = (\frac{\theta°}{360°})(\pi d)$
   $C(\text{arc}) = (\frac{32°}{360°})(\pi 16)$
   $C(\text{arc}) = 4.5$ inches
3. To calculate, find the area of the entire garden and then subtract the area of the sector:
   $A(\text{whole}) - A(\text{sector}) = (\pi r^2) - (\frac{\theta°}{360°})(\pi r^2)$
   $A(\text{whole}) - A(\text{sector}) = (\pi 11^2) - (\frac{140°}{360°})(\pi 11^2)$
   $A(\text{whole}) - A(\text{sector}) = (380) - (148)$
   $A(\text{whole}) - A(\text{sector}) = 232$ ft.^{2}

real-world problems with area and perimeter

We only think when confronted with a problem.
—JOHN DEWEY

In this lesson you will learn how to calculate the missing side lengths in polygons when you know the area or perimeter. You will also calculate the areas of more complex figures.

USING AREA TO FIND MISSING INFORMATION

In some situations, the area of a polygon will be given, and you will need to solve for its width or length. To do so, you will need to choose the correct formula, plug in the given information, and then solve for the missing dimension. Sometimes, you will use a formula for perimeter, and other times it will be appropriate to use a formula for area.

Example: Marni is painting the walls of her house. One gallon of paint covers 400 ft.^{2} of wall space with a single coat. If Marni's walls are 8 feet tall, how many horizontal feet of wall space can she cover with a single coat of paint?

Solution: Painting walls in a house involves knowing about the space *inside* rectangles, so plug the given information into the formula for the area of the rectangles. Then solve for the missing dimension:

$A = \text{length} \times \text{width}$

$400 = \text{length} \times (8)$

$\text{Length} = 50$

Marni can cover 50 horizontal feet of wall space with one gallon of paint.

PRACTICE 1

1. A horse is tethered with a rope to a stake in an open pasture. If the largest circle the horse can walk around the stake is 94.2 meters, how many meters long is the rope that the horse is tied with?

2. Ryan has a package of zinnia seeds. The instructions say that there are enough seeds to cover 50 ft.^{2} of ground. He wants to make a flower box in front of his living room window that is 12.5 feet long. How wide should the box be?

3. Jocelyn purchased 16 feet of decorative trim to put around the edge of a rectangular painting she will make over the weekend. What are three different possible combinations for the length and width of the painting, if she wants to use all of the trim she purchased?

4. Using the information in problem **3**, Jocelyn decides to make a circular painting and still wants to use all 16 feet of the decorative trim around the outside edge. What will the diameter of this circular painting be?

5. A square piece of fabric has an area of 110.25 in.^{2}. What is the perimeter of this piece of fabric?

FINDING THE AREA BETWEEN SHAPES

Another common type of problem relating to area is when one shape is contained within another shape. You will encounter this situation often in the real world. In order to solve this type of problem, it is necessary to find the area of the larger figure, and then subtract from that measure the area of the smaller figure. Look at the following figure as you read through the following example.

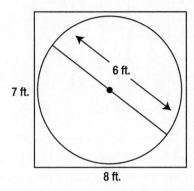

7 ft.

6 ft.

8 ft.

Example: A circular fountain sits in a rectangular garden space. The garden space is 8 feet long by 7 feet wide and the fountain has a diameter of 6 feet. What is the area, to the nearest tenth, of the garden space that will not be taken up by the fountain?

Solution: The area of the garden space will be $A = lw = (7)(8) = 56$ ft.^{2}. The area of the circular fountain will be $A = \pi r^2 = \pi(3^2) \approx 28.26$ ft.^{2}. Therefore, the area of the garden space that is not covered by the fountain is the difference of the two shapes: $56 - 28.3 = 27.7$ ft.^{2}.

TIP: To find the area between two different figures, subtract the area of the smaller figure from the area of the larger figure.

One thing to be aware of when you're working with real-world problems is that the units may not always be the same. For example, if you are given one dimension in inches and one in feet, it is necessary to convert both dimensions into feet, or both into inches, before attempting to solve the problem.

PRACTICE 2

1. In their rectangular backyard, the Spieth family has a fire pit as shown in the following figure. The fire pit is 8 feet long by 3 feet wide. They want to fill a 14-foot by 11-foot area surrounding the fire pit with concrete. How many square feet will the concrete section surrounding the fire pit contain?

2. A penny is placed on top of a quarter. If the penny has a diameter of 19 mm and the quarter has a diameter of 24 mm, what is the area to the nearest millimeter of the top surface of the quarter that is not covered by the penny?

3. A bathroom counter has two circular sinks as shown in the following figure. The counter is 5 feet by 2 feet in size and each sink is $1\frac{1}{4}$ feet in diameter. How many square feet of tiles will Kerry need to cover the counter?

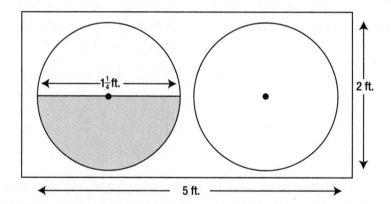

4. Tivoli makes a flag to bring to the beach so that her friends can find her. She first uses a piece of teal fabric that is 5 feet long by 3 feet wide. She then adds to it three pieces of printed material: two triangles and one semicircle with the dimensions given in the next figure. What is the area in square feet of the original teal fabric that will be visible in her flag?

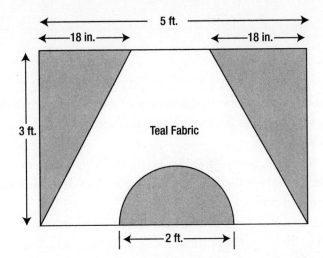

5. Jago is using tiles that are 3 inches long by 2 inches wide to cover a kitchen counter. If the counter is 6 feet long by 2.5 feet wide, how many tiles will he need to cover the counter completely? (*Hint:* Find the area of the counter *in inches*, and divide it by the area in square inches of each tile.)

ANSWERS

Practice 1

1. This problem involves the circumference of a circle, since it is discussing the distance around a stake that a horse can walk.
   $C = 2\pi r$ so $94.2 = 2\pi r$
   $2\pi r \approx 2(3.14)r \approx 94.2$, so $r \approx \frac{94.2}{6.28} \approx 15$ meters. The tether is 15 m long.

2. This problem involves the area of a rectangle, since it is discussing the space within a rectangular flower box.
   $A = lw = 12.5w = 50$, so $w = 4$. The box should be 4 feet wide.

3. This problem involves the perimeter of a rectangle, which uses the formula $P = 2l + 2w$ or $P = 2(l + w)$. Since the perimeter is 16 feet, $16 = 2(l + w)$ or $8 = l + w$. Therefore, any pairing of length and width that sum to 8 would work as a correct answer. Some possibilities are $l = 7$ ft., $w = 1$ ft.; $l = 6$ ft., $w = 2$ ft.; and $l = 5$ ft., $w = 3$ ft. Even $l = 4$ ft., $w = 4$ ft. would work since although this is a square, by definition, a square is a rectangle.

4. This problem involves the circumference, since it is investigating how large a circle 16 feet of ribbon can go around.
   $C = 2\pi r$, so $16 = 2\pi r$
   $2\pi r \approx 2(3.14)r \approx 16$, so $r \approx \frac{16}{6.28} \approx 2.5$ feet. The diameter of the painting could be ≈ 5 feet.

5. Since the formula for the area of a square is $A = s^2$, you can solve for s in the equation $110.25 = s^2$. In this case, the side length will be 10.5. The perimeter of a square is $4s$, so $4(10.5) = 42$ inches.

Practice 2

1. Since all given dimensions are in feet and both figures are rectangles, just use the formula $A = LW$ for each shape and find the difference:

   $A(\text{whole}) - A(\text{fire pit}) = (L)(W) - (l)(w)$

   $A(\text{whole}) - A(\text{fire pit}) = (14)(11) - (8)(3)$

   $A(\text{whole}) - A(\text{fire pit}) = (154) - (24)$

   $A(\text{whole}) - A(\text{sector}) = 130 \text{ ft.}^2$

2. Since both dimensions are in millimeters (mm) you will not need to do any conversions. Do not forget to divide each diameter by 2 in order to get the radius to use in the formula $A = \pi r^2$. Find the area for each coin and subtract their areas:

   $A(\text{quarter}) - A(\text{penny}) = \pi r^2 - \pi r^2$

   $A(\text{quarter}) - A(\text{penny}) = \pi 12^2 - \pi 9.5^2$

   $A(\text{quarter}) - A(\text{penny}) = (452.16\pi) - (283.385\pi) \approx 168.775 \text{ mm}^2$

3. The area of the entire counter will be $5 \times 2 = 10 \text{ ft.}^2$. The radius of the sinks is $\frac{1.25}{2} = 0.625$ feet. The combined area of the two sinks will be $2(A) = 2(\pi r^2) = 2(\pi)(0.625^2) \approx 2.5 \text{ ft.}^2$. The difference of these two areas is 7.5 ft.^2.

4. The area of the teal fabric that Tivoli begins with is $5 \times 3 = 15 \text{ ft.}^2$. The base of each of the triangles is 3 feet and their heights are each 1.5 feet ($\frac{18 \text{ inches}}{12 \text{ inches}} = 1.5$ feet). Therefore, the area of these two triangles is $2(A) = 2(\frac{1}{2})(\text{base})(\text{height}) = (1)(3)(1.5) = 4.5 \text{ ft.}^2$. The area of the semicircle with a radius of 1 is $(\frac{1}{2})(A) = (\frac{1}{2})(\pi r^2) = (\frac{1}{2})(\pi 1^2) = 1.57 \text{ ft.}^2$. Subtracting the circular and triangular material from the teal gives $15 - 4.5 - 1.57 = 8.93 \text{ ft.}^2$.

5. The area of the counter is $6 \times 2.5 = 15 \text{ ft.}^2$. Every square foot has an area of 144 square inches, since a square foot is made up of 12 inches by 12 inches. Therefore, for each of the 15 square feet of counter, there are 144 square inches, so we can calculate that there are 15×144 or 2,160 square inches of area within the counter. Each tile is $2 \times 3 = 6 \text{ in.}^2$. Last, to see how many 6-square-inch tiles are needed to cover 2,160 square inches, divide the two: $\frac{2,160}{6} = 360$ tiles.

surface area

Since the mathematicians have invaded the theory of
relativity, I do not understand it myself any more.
—ALBERT EINSTEIN

In this lesson you will first learn how to identify three-dimensional objects.
You will then learn about surface area and how to calculate it for prisms and
cylinders.

IDENTIFYING PRISMS

Although the word "prism" is something that you may not have heard until
today, you have been looking at prisms all day. A **prism** is a three-dimensional
figure that has two congruent polygon ends and parallelograms as the remaining sides. A pizza box is a prism that you are probably familiar with. The sides
of prisms can all be called **faces**. The congruent ends are more specifically
referred to as **bases**. The remaining sides, which are parallelograms, are the **lateral faces**. The *shape of the polygon base* is what determines the name of a prism
as shown in the next figure.

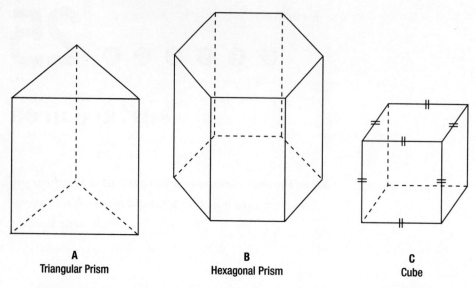

A
Triangular Prism

B
Hexagonal Prism

C
Cube

Prism A has two parallel triangular bases, so it is a **triangular prism**. Prism B has parallel hexagons as bases, so it is a **hexagonal prism**. Prism C has a square base and the lateral faces are also all squares. Instead of being called a "square prism," this figure is called a **cube**. Notice that in all three prisms, the lateral faces are rectangles. **Edges** in prisms are the segments formed where two faces come together. Edges that are parallel always have the same length. **Vertices** are the corners where three edges come together.

TIP: A *prism* is a three-dimensional figure with two congruent and parallel polygon bases and parallelogram sides. Prisms are named after their polygon bases.

PRACTICE 1

1. How many faces in total does a rectangular prism have?

2. How many faces in total does an octagonal prism have?

3. How many edges does a triangular prism have?

4. How many vertices does a triangular prism have?

For questions 5–8, name the following prisms.

5.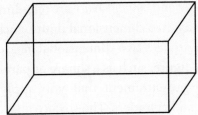

6.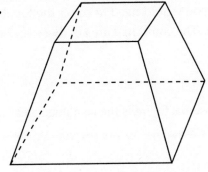

7.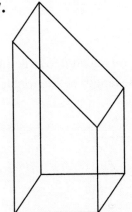

8.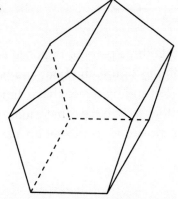

WHAT IS SURFACE AREA?

In previous lessons you learned that *perimeter* and *circumference* are one-dimensional measurements of the distance around a two-dimensional figure, such as a square or rectangle. You also learned that *area* is a two-dimensional measurement of the space *inside* a two-dimensional figure, such as a square or parallelogram. **Surface area** is a third type of measurement that you apply to three-dimensional figures, such as prisms and cylinders. The surface area of a 3D figure is the combined area of all the figure's faces. For example, if you needed to wrap a birthday gift in a box, you would need to cover the top, bottom, front, back, and two sides of the box. The combined area of these six sides is called the *surface area* of the box.

....

TIP: The *surface area* of a three-dimensional figure is the combined area of all the figure's outward-facing surfaces.

....

SURFACE AREA OF RECTANGULAR PRISMS

A rectangular prism has six faces. Its three dimensions are referred to as length, width, and height. In the next figure, you can see that the length of the front face is 9 and its height is 4. The width of the prism is 4 units.

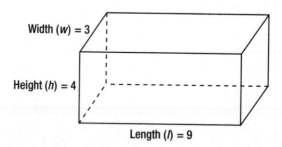

The six faces of rectangular prisms include three pairs of opposite congruent faces that have the same dimensions. In this figure, the front and back faces have dimensions of 9 units by 4 units; the left and right faces have dimensions of 4 units by 3 units; and the top and bottom faces have dimensions of 9 units by 3 units. To find the surface area of this prism, we could set up a table as in the next figure.

Faces	Formula for Area	Area of One Face
Front and back	$L \times H$	$9 \times 4 = 36$
Left and right	$H \times W$	$4 \times 3 = 12$
Top and bottom	$L \times H$	$9 \times 3 = 27$

Subtotal : $36 + 12 + 27 = 75$ units
Multiplied by 2 since there are 2 faces with each
dimension: $(75)(2) = 150$ units = Surface Area

The area of each pair of faces is the product of two different dimensions. The three distinct products have a sum of 75 units. You then double this subtotal, since there are two identical faces with each given area. The result is a total surface area of 150 units^2. This would be a lot of work to go through every time you wanted to find the surface area (SA) of a rectangular prism.

Therefore, consider the formula, $SA = 2lh + 2hw + 2lw$. In this formula, the three different products, lh, hw, and lw, represent the three areas of *one* of the distinct front, top, and left faces. Each of these areas is multiplied by 2, and this sum is the total surface area. This formula can be simplified by factoring a 2 out of the terms and writing it as $SA = 2(lh + hw + lw)$.

....

TIP: The surface area of a rectangular prism is: $SA = 2(lh + hw + lw)$.

....

SURFACE AREA OF CUBES

The surface area of a cube is an easier formula to remember.

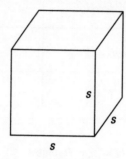

In this figure you can see that all of the edges of the cube are labeled *s*. In a cube, each of the six faces has an area of s^2, since the length, width, and height are all *s* units long. Since there are six faces, each with an area of s^2, the formula for the surface area for a cube is $SA = 6s^2$. When you're using this formula, remember to follow the proper order of operations by squaring the side length *first*, before multiplying by 6.

··

TIP: The surface area of a cube is: $SA = 6s^2$.

··

PRACTICE 2

Find the surface area of each rectangular prism or cube.

1.

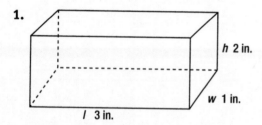

h 2 in.

w 1 in.

l 3 in.

2.

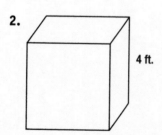

4 ft.

3.

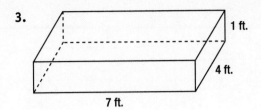

1 ft.

4 ft.

7 ft.

4.

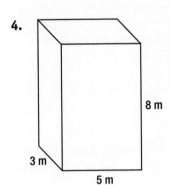

8 m

3 m

5 m

5.

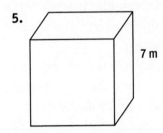

7 m

6. The surface area of a rectangular prism is 288 in.^{2}. If the length of the front face is 12 inches and the height of the front face is 6 inches, what is the width of the rectangular prism?

7. The surface area of a cube is 486 ft.^{2}. Find the length of each edge.

SURFACE AREA OF CYLINDERS

To understand the method for finding the surface area for a cylinder, imagine doing the following exercise with the short cardboard cylinder pictured next.

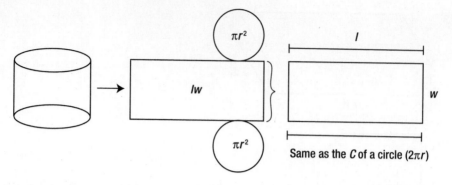

In the leftmost illustration, you can see that there are two parallel circular bases on the top and bottom of the cylinder. In the center illustration, the cardboard cylinder has been cut and unrolled, so that the cylindrical part is now a flattened rectangle with an area of *lw*. The areas of the circular top and bottom can each be found by using $A = \pi r^2$. In the rightmost illustration, it is shown that the *length* of the flattened rectangle is actually equal to the circumference of the circular bases. The *width* of the rectangle section of the cylinder is actually the height of the cylinder. Therefore, the formula (*circumference*)(*height*) is used to identify the area of the rectangular section of the cylinder.

In summary, the surface area of a cylinder is equal to: (area of two circular bases) + (area of the rectangular portion). The area of the rectangular portion is found by multiplying the *circumference* of the circular base by the *height* of the cylinder.

· ·

TIP: The surface area of a cylinder is:
$$SA = 2\pi r^2 + (2\pi r)(h)$$

· ·

PRACTICE 3

Find the surface area of each of the cylinders that are drawn or described next. Use 3.14 for π and round your answers to the nearest tenth.

1. 1 m

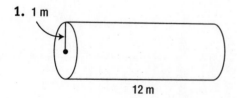

12 m

2.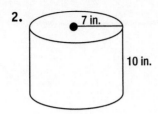
7 in.

10 in.

3. Cylinder: $r = 6$ ft., $h = 16$ ft.

4. Cylinder: $d = 18$ cm, $h = 3$ cm

ANSWERS

Practice 1

1. A rectangular prism has 6 faces.
2. An octagonal prism has 10 faces.
3. A triangular prism has 9 edges.
4. A triangular prism has 6 vertices.
5. Rectangular prism
6. Trapezoidal prism
7. Trapezoidal prism (the bases are the front-facing and back-facing trapezoids)
8. Pentagonal prism

Practice 2

1. $SA = 2(lh + hw + lw) = 2(3 \times 2 + 2 \times 1 + 3 \times 1) = 2(11) = 22$ in.^{2}
2. $SA = 6s^2 = 6(4)(4) = 96$ in.^{2}
3. $SA = 2(lh + hw + lw) = 2(7 \times 1 + 4 \times 1 + 7 \times 4) = 2(39) = 78$ ft.^{2}
4. $SA = 2(lh + hw + lw) = 2(5 \times 8 + 5 \times 3 + 3 \times 8) = 2(79) = 158$ m^2
5. $SA = 6s^2 = 6(7)(7) = 294$ in.^{2}
6. $SA = 2(lh + hw + lw)$
   $288 = 2(12 \times 6 + 6w + 12w)$
   $288 = 2(72 + 18w)$
   $288 = 144 + 36w$
   $144 = 36w$
   $w = 4$ inches

7. $SA = 6s^2$

    $486 = 6s^2$

    $81 = s^2$

    $s = 9$ feet

Practice 3

1. $SA = 2\pi r^2 + (2\pi r)(h) = 2(3.14)1^2 + (2)(3.14)(1)(12) = 6.28 + 6.28(12) = 81.6 \text{ m}^2$

2. $SA = 2\pi r^2 + (2\pi r)(h) = 2(3.14)(7^2) + (2)(3.14)(7)(10) = 307.72 + 439.6 =$
747.3 in.^{2}

3. $SA = 2\pi r^2 + (2\pi r)(h) = 2(3.14)(6^2) + (2)(3.14)(6)(16) = 226.08 + 602.88 =$
829 ft.^{2}

4. $SA = 2\pi r^2 + (2\pi r)(h) = 2(3.14)(9^2) + (2)(3.14)(9)(3) = 508.68 + 169.56 =$
678.2 cm^2

volume

*Creative mathematicians now, as in the past, are inspired
by the art of mathematics rather than by
any prospect of ultimate usefulness.*

—ERIC TEMPLE BELL

In this lesson you will learn about volume and how to calculate the volume for
a variety of figures.

WHAT IS VOLUME?

"How much volume would you like?" This really should be the question that
you are asked when you're ordering a glass of juice in a restaurant. It is not really
the *size* of the glass that you are interested in, but the *amount* of juice that goes
into that glass. **Volume** refers to the amount of space occupied inside a three-
dimensional object. Similarly to how area is measured in squares, volume is
measured in cubes. Think about neatly arranging one-inch cubes, side-by-side,
in a small shoebox. The total number of cubes that would fill the box would be
the volume of the box in cubic inches.

TIP: *Volume* is the amount of space enclosed by a three-dimensional
object.

PRACTICE 1

For questions **1–6**, state whether the concept of perimeter, area, surface area, or volume would be used to investigate the real-world situation that Dani is in.

1. Dani needs to paint the outside of a box black so that it can be used as a prop on stage for the school play.

2. Dani needs to buy the right-sized air-conditioner to cool down her studio apartment.

3. Dani was late for soccer practice and had to run 5 full laps around the field.

4. Dani wonders how much it will cost to fill the pool with water if each gallon from her hose costs $0.07.

5. Dani is buying fabric to make a tablecloth.

6. Dani wants to put tile around the window in her living room.

FINDING THE VOLUME OF RECTANGULAR PRISMS

To understand how volume is calculated, read on while looking at the rectangular prism in the next figure.

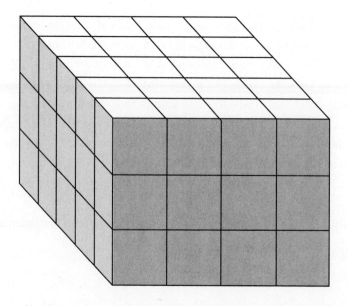

The face of the rectangular prism, which is shaded the darkest, has a length of 4 units and a height of 3 units. Therefore the area of the front face is $4 \times 3 = 12$ square units. The width of this rectangular prism is 5 units. Notice that for each of the 12 cubes in the front face there is a row of 5 cubes extending toward the back of the prism. Knowing this, you can find the volume by multiplying the number of squares in the front face by the number of cubes behind it. Therefore, the number of cubes that make up this prism is $12 \times 5 = 60$, and its volume is 60 cubic units, which is written 60 units^3. We just found the **volume** of this **rectangular prism** by multiplying the **(length)(width)(height)**. (Volume is always expressed with an exponent of 3 after the units, to represent the three different dimensions it measures.)

TIP: Volume of a rectangular prism = *(length)(width)(height)*

FINDING THE VOLUME OF CUBES

In cubes, the length, width, and height are all the same length, *s*. Therefore, the formula for finding the **volume of a cube** is *(side)(side)(side)* or **(side)**^{3}.

TIP: Volume of a cube = s^3.

Refer to the next figure while you look through the following two examples.

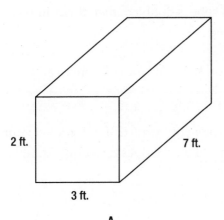

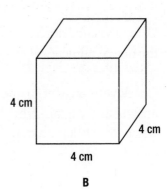

A	B

Example 1: Find the volume of the rectangular prism in Figure A.

Solution: $V = (length)(width)(height) = (2)(3)(7) = 42$ ft.^{3}.

Example 2: Find the volume of the cube in Figure B.

Solution: $V = s^3 = 4^3 = 64$ cm^3.

PRACTICE 2

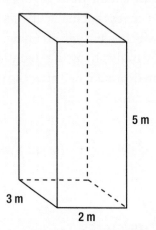

5 m

3 m

2 m

1. Find the volume of the rectangular prism in this figure.

2. Consider the prism in question **1**. If another prism with the same base dimensions of 2 meters by 3 meters, but with a height of 7 m, was compared to the original prism, how many more cubic meters does the new prism contain?

3. What is the volume of a cube with a side length of 3 yards?

4. True or false: If the side length of a cube is doubled, then the volume of the new cube would also be doubled.

5. Using the cube described in question **3**, double the side length and find the new volume.

6. Check your answer to question **4**. Were you correct? If not, correct your answer and explain why the volume of a cube does not simply double when its side length doubles.

FINDING THE VOLUME OF CYLINDERS AND TRIANGULAR PRISMS

Given *any* right prism, if you can calculate the area of one of its congruent bases, then multiplying that area by the height of the prism will give you the volume. To understand this, look at the following figure.

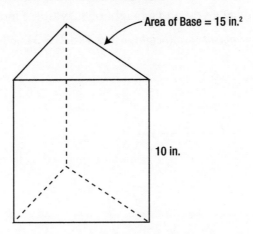

Area of Base = 15 in.²

10 in.

If you know that the area of the base is 15 in.², then you could multiply that by 10 inches to get the total volume of 150 in.³.

....

TIP: Volume of any prism = (*area of base*)(*height*)

....

This formula is useful because it can be applied to cylinders and other various prisms. You can even use it for rectangular prisms or cubes if you forget those formulas.

Refer to the next figure while you look through the following two examples.

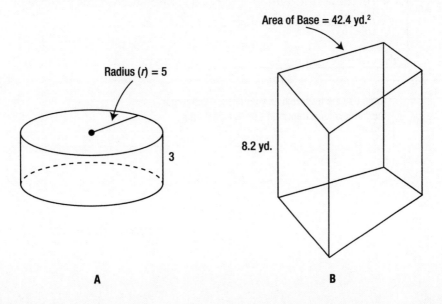

Area of Base = 42.4 yd.²

Radius (*r*) = 5

3

8.2 yd.

A

B

Example 1: Find the volume of the cylinder in Figure A.

Solution: $V = (area\ of\ base)(height)$. In this case, the area of the base is πr^2 since the base is a circle. $V = (\pi r^2)(h) = (\pi 5^2)(3) = 235.5\ m^3$.

Example 2: Figure B contains a base that is an irregularly shaped prism with an area of 42.4 yd.^{2} and a height of 8.2 yards. Find its volume.

Solution: $V = (area\ of\ base)(height) = (42.4)(8.2) = 347.68\ yd.^3$.

PRACTICE 3

1.

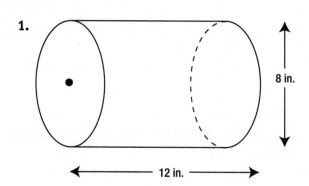

8 in.

12 in.

2.

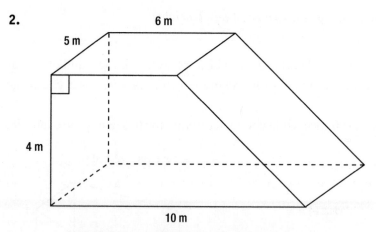

6 m

5 m

4 m

10 m

(*Hint:* Use the trapezoid as your base. If you forgot the formula for a trapezoid's area, look it up.)

3.

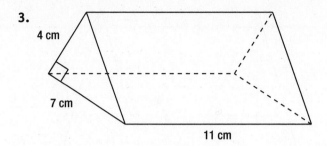

Hint: Use the triangle as your base. You should remember the formula for the area of a triangle!

4.

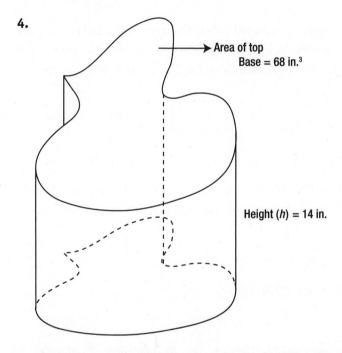

Area of top
Base = 68 in.^{3}

Height (*h*) = 14 in.

Can the formula given be used for this organically shaped vase? It has a curvy top surface, but the sides are perfectly straight. If the volume can be identified, find it. If it can't be calculated, explain why.

FINDING THE VOLUME OF PYRAMIDS, CONES, AND SPHERES

Last, we will look at how to calculate the volume of pyramids, cones, and spheres. A **pyramid** is a three-dimensional shape with a regular polygon base and congruent triangular sides that meet at a common point. Pyramids are named by their base: Illustration A is a triangular pyramid, illustration B is a rectangular pyramid, and illustration C is a hexagonal pyramid.

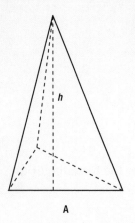
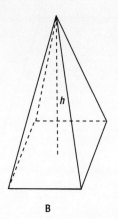
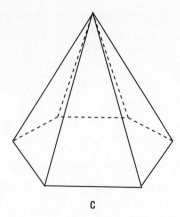

A   B   C

As long as you are given, or can calculate, the area of the base of a pyramid, and you know the perpendicular height from the base to the vertex, you can determine the **volume of a pyramid** by using $V = \frac{1}{3}(area\ of\ base)(height)$.

TIP: Volume of Pyramid = $(\frac{1}{3})(area\ of\ base)(height)$.

The **volume of a cone** is also $\frac{1}{3}$ the product of its base area and height. However, in the case of cones, the area of the base is the formula for the area of a circle.

TIP: Volume of cone = $(\frac{1}{3})(\pi r^2)(height)$.

Spheres have only one measurement to use in formulas—their radii. In order to calculate the **volume of a sphere**, use the following formula.

TIP: Volume of sphere = $(\frac{4}{3})\pi r^3$.

PRACTICE 4

1.

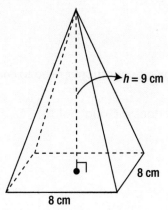

$h = 9$ cm

8 cm

8 cm

2.

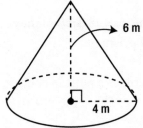

6 m

4 m

3.

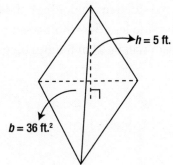

$h = 5$ ft.

$b = 36$ ft.²

Note, the area of the base is given as b.

4.

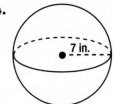

7 in.

ANSWERS

Practice 1

1. Surface area
2. Volume—how many cubic feet of air does the air-conditioner need to cool?
3. Perimeter
4. Volume—how many cubic gallons of water does the pool hold?
5. Area
6. Perimeter

Practice 2

1. $V = (length)(width)(height) = (2)(3)(5) = 30 \text{ m}^3$.
2. $V = (length)(width)(height) = (2)(3)(7) = 42 \text{ m}^3$. The new prism will have $42 - 30 = 12$ more cubic meters than the original prism.
3. $V = s^3 = 3^3 = 27 \text{ yd.}^3$.
4. False: $V = s^3$. Since the side length is doubled, replace s with $2s$ and use the formula for volume again: $V = (2s)^3 = 2^3 s^3 = 8s^3$. Therefore, the new volume will be 8 times the original volume.
5. $V = s^3 = 6^3 = 216 \text{ yd.}^3$.
6. It is a common mistake to think that when the side length doubles, the volume will also double. However, since the side length is multiplied by itself three times, the volume of the new prism will be 2^3 or 8 times bigger.

Practice 3

1. $V = $ (area of circular base)(prism height)
$V = (\pi r^2)($ prism height)
$V = (\pi 4^2)(12) = 192\pi$ in.^{3} or 602.88 in.^{3}.

2. $V = $ (area of trapezoidal base)(prism height)
$V = [(\frac{1}{2})(h)(b_1 + b_2)]($ prism height)
$V = (\frac{1}{2})(4)(10 + 6)(5) = 80$ m^3

3. $V = $ (area of triangular base)(prism height)
$V = [(\frac{1}{2})(bh)]($ prism height)
$V = [(\frac{1}{2})(4)(7)](11) = 154$ cm^3

4. Since the area of the irregular top of the vase is given, and since its sides are straight, we can use the formula, (area of base)(height), to calculate the volume of the vase: $68 \times 14 = 952$ in.^{3}.

Practice 4

1. Volume of pyramid $= (\frac{1}{3})$(area of base)(height)
$V = (\frac{1}{3})(8 \times 8)(9) = 192$ cm^3

2. Volume of cone $= (\frac{1}{3})(\pi r^2)$(height)
$V = (\frac{1}{3})(\pi 4^2)(6) = 32\pi$ m^3 or 100.48 m^3

3. Volume of pyramid $= (\frac{1}{3})$(area of base)(height)
$V = (\frac{1}{3})(36)(5) = 60$ ft.^{3}

4. Volume of sphere $= (\frac{4}{3})\pi r^3 = (\frac{4}{3})(\pi)(7^3) = 457.33\pi$ in.^{3} or 1,436 in.^{3}

coordinate geometry

The shortest distance between two points is under construction.
—BILL SANDERSON

In this lesson you will first learn about points on a coordinate plane. You will learn how to plot coordinate pairs and how to find the midpoint and distance between any two points.

WHAT IS THE COORDINATE PLANE?

A **coordinate plane** is similar to a city map, except it is a special type of map used in mathematics to plot points, lines, and shapes. The coordinate plane is made up of two main axes that intersect at the **origin**, to form a right angle. The *x*-axis is the main horizontal axis, and the *y*-axis is the main vertical axis. These perpendicular axes divide the coordinate plane into four sections, or **quadrants**. The quadrants are numbered in counterclockwise order, beginning with the top-right quadrant. See the following figure to understand all the terms that have just been presented.

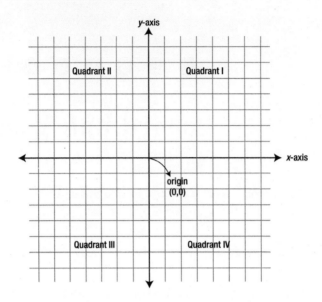

In order to give a friend directions, you need to know from where she is starting. In a coordinate plane, the starting point is the **origin**, or center point. (This makes sense, since "origin" means the place where something begins.) The origin is located at (0,0). Both the *x*- and *y*-axes are numbered like number lines. Starting from the origin, positive movements are moving right or up and negative movements include moving left or down.

....

TIP: The *origin* is the center point of a coordinate plane. It is located at (0,0), where the *x*-axis and *y*-axis intersect.

....

PLOTTING POINTS IN THE COORDINATE PLANE

Points in the coordinate plane have an *x*-coordinate and a *y*-coordinate that show where the point is in relation to the origin. Points are presented in the form (*x,y*). The first number represents the point's *horizontal* distance and direction from the origin. The second number represents the point's *vertical* distance and direction from the origin.

....

TIP: A *coordinate pair* is a point on a coordinate graph. It is an ordered pair of numbers written as (*x,y*).

....

For example, point *A* in the next figure is located at (–3,5). Notice that this point is 3 units to the *left* of the origin and 5 units *above* it. Similarly, point *B* in the figure at (2,–4) represents a point that is 2 units to the *right* of the origin and 4 units *below* it. When plotting points have 0 in them, the point will sit on one of the axes since either its vertical or its horizontal movement will be zero units.

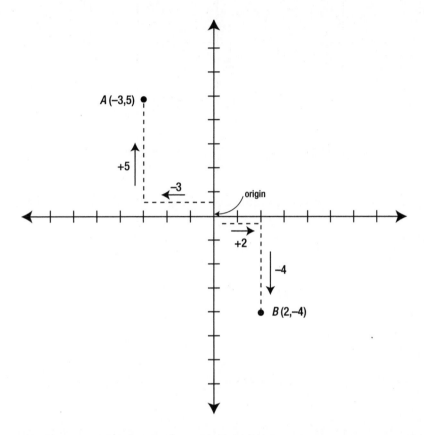

Since the *x*-coordinate always comes first in coordinate pairs, it is necessary that your first movement from the origin be horizontal. To plot a point, start at the origin, and move *x* units to the left if the *x*-coordinate is negative, or *x* units to the right if it is positive. Next, look at the *y*-coordinate. Move *y* units up if it is positive, or *y* units down if it is negative.

PRACTICE 1

For questions **1–6**, graph and label the following points on the given coordinate plane.

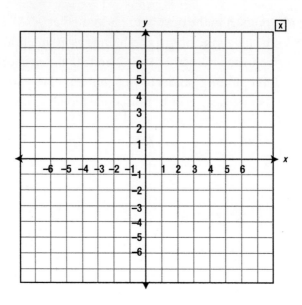

1. A (–3,5)

2. B (5,–3)

3. C (0,4)

4. D (2.5,6)

5. E (–4.5,–6)

6. F (5,0)

For questions **7–12**, identify what quadrant the point lies in. If the point is not in a quadrant, determine what axis it sits on.

7. (–2,5)

8. (1,–8)

9. (0,3.75)

10. (12,16)

11. (–7,–7)

12. (9,0)

For questions **13–20**, write the coordinate for each point.

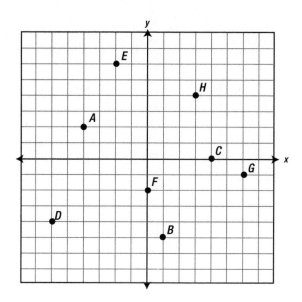

13. *A*

14. *B*

15. *C*

16. *D*

17. *E*

18. *F*

19. *G*

20. *H*

FINDING THE DISTANCE BETWEEN TWO POINTS

The distance between two points in a coordinate plane is the shortest route between the two points. Unless the two points have the same *x*- or *y*-coordinate, the distance will be represented by a diagonal line.

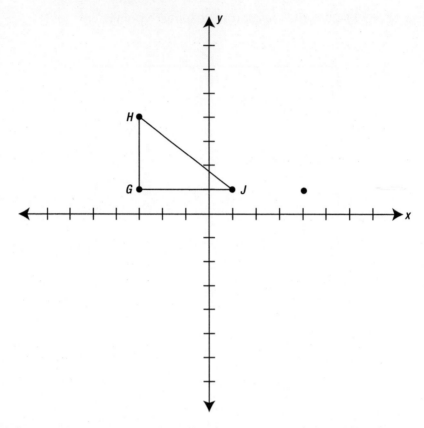

Looking at this figure, it is easy to count the vertical distance between points *G* and *H* as 3 units. You can also count the horizontal distance between points *G* and *J* as being 4 units. What is not easy to count is the diagonal distance between points *H* and *J*. However, since connecting these three points makes a right triangle, we can use the Pythagorean theorem to calculate the distance between *H* and *J*.

$$(\overline{HG})^2 + (\overline{GJ})^2 = (\overline{HJ})^2$$

$$(3)^2 + (4)^2 = (\overline{HJ})^2$$

$$9 + 16 = (\overline{HJ})^2$$

$$25 = (\overline{HJ})^2$$

$$\overline{HJ} = 5$$

It would be a lot of work to have to plot your points, connect them, and use the Pythagorean theorem every time you needed to find the distance between two points! Let us figure out a formula to use by taking a look at the next figure.

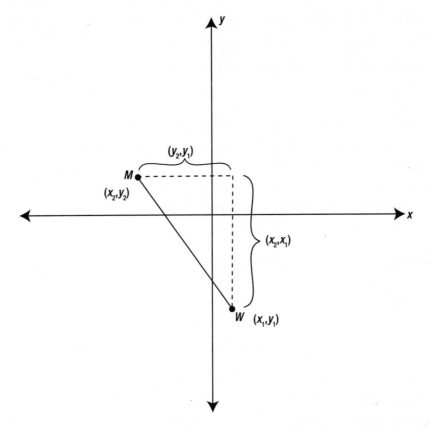

In this figure, point W is labeled (x_1, y_1) and point M is labeled (x_2, y_2). The distance we are trying to solve for is the *solid* line between points M and W. The *dotted vertical* line to the right represents the vertical distance between the points, which can be calculated with $(y_2 - y_1)$. The dotted horizontal line up top represents the horizontal distance between the points, which can be calculated with $(x_2 - x_1)$. Since $(x_2 - x_1)$ and $(y_2 - y_1)$ represent the length of two legs of a right triangle, we can use these algebraic expressions in the Pythagorean theorem in order to come up with a formula for distance.

$$(\overline{MW})^2 = (x_2 - x_1)^2 + (y_2 - y_1)^2$$

In order to solve for the length of $\overline{MW}$ we need to take the square root of both sides of this equation:

$$\sqrt{(\overline{MW})^2} = \sqrt{((x_2 - x_1)^2 + (y_2 - y_1)^2}$$
$$\overline{MW} = \sqrt{(x_2 - x_1)^2 + (y_2 - y_1)^2}$$

You can generalize this discovery to find the distance between any two points in the coordinate plane.

∙∙∙

TIP: The *distance, d,* between any two points $A(x_1,y_1)$ and $B(x_2,y_2)$ in the coordinate plane is $d = \sqrt{(x_2 - x_1)^2 + (y_2 - y_1)^2}$

∙∙∙

When you're looking for the distance between two points, it is not important which point you chose to be your $A(x_1,y_1)$ and which point to be your $B(x_2,y_2)$. In fact the order in which you use x_z and x_x is not important since when the difference of these two numbers is squared, the result will always be positive. When you're working with the distance formula, it is important to remember that the square root of a sum of two terms is *not* equal to the sum of the individual square roots of those terms: $\sqrt{a^2 + b^2} \neq \sqrt{a^2} + \sqrt{b^2}$. Therefore, it is necessary that you always *first* find the sum of $(x_2 - x_1)^2 + (y_2 - y_1)^2$ *before* taking the square root of the sum.

Example: Find the distance between the points (3,8) and (–5,12).

Solution: Let (3,8) be (x_1,y_1); and let (–5,12) be (x_2,y_2).

$$d = \sqrt{(x_2 - x_1)^2 + (y_2 - y_1)^2}$$
$$d = \sqrt{(-5 - 3)^2 + (12 - 8)^2}$$
$$d = \sqrt{(-8)^2 + (4)^2}$$
$$d = \sqrt{64 + 16}$$
$$d = \sqrt{80}$$
$$d = 8.9$$

PRACTICE 2

1. The distance formula $d = \sqrt{(x_2 - x_1)^2 + (y_2 - y_1)^2}$, comes from which theorem?

2. True or false: $d = \sqrt{(x_2 - x_1)^2 + (y_2 - y_1)^2} = \sqrt{(x_2 - x_1)^2} + \sqrt{(y_2 - y_1)^2}$

3. True or false: The distance between two points can always be counted accurately as long as the points are plotted on graph paper.

For questions 4–7, calculate the distance between the given coordinates.

4. (3,5) and (–1,5)

5. (3,–2) and (–5,4)

6. (–3,–8) and (–6,–4)

7. (–1,–6) and (4,6)

FINDING THE MIDPOINT BETWEEN TWO POINTS

The midpoint of two coordinate pairs is the point that is equal in distance from both of the coordinate pairs. Since taking the average of two numbers finds the middle number, it makes sense that the **midpoint formula** involves finding the average. Consider again points A, at (x_1,y_1), and B, at (x_2,y_2). The x-coordinate that will be a midpoint, equidistant from x_1 and x_2, will be the average of these coordinates: $\frac{x_1+x_2}{2}$. The same goes for finding the y-coordinate, which will be equidistant from both y_1 and y_2: $\frac{y_1+y_2}{2}$. These two expressions are combined to create the midpoint formula.

> **TIP:** The *midpoint* between any two points $A(x_1,y_1)$ and $B(x_2,y_2)$ is $\left(\frac{x_1+x_2}{2}, \frac{y_1+y_2}{2}\right)$

Example: Find the midpoint between the points (3,8) and (–5,12).

Solution: Let (3, 8) be (x_1,y_1), and let (–5,12) be (x_2,y_2).

$$\text{midpoint} = \left(\frac{x_1+x_2}{2}, \frac{y_1+y_2}{2}\right)$$

$$\text{midpoint} = \left(\frac{3+-5}{2}, \frac{8+12}{2}\right)$$

$$\text{midpoint} = \left(\frac{-2}{2}, \frac{20}{2}\right)$$

$$\text{midpoint} = (-1,10)$$

Sometimes, the midpoint and one of the endpoints will be given, and you will need to find the coordinates of the other endpoint.

Example: Find the coordinates of the endpoint of a line segment that has one endpoint at (11,6) and a midpoint at (7,–4).

Solution:

$$\text{midpoint} = \left(\frac{x_1 + x_2}{2}, \frac{y_1 + y_2}{2} \right)$$

$$(7,-4) = \left(\frac{11 + x_2}{2}, \frac{6 + y_2}{2} \right)$$

$(7,-4)$ must equal $\left(\frac{14}{2}, \frac{-8}{2} \right)$

So working backwards you can reason:

$$(7,-4) = \left(\frac{11 + 3}{2}, \frac{6 + -14}{2} \right)$$

$$\text{endpoint} = (3,-14)$$

PRACTICE 3

Find the midpoint between the coordinate pairs for questions **1–5**.

1. (2,5) and (–10,11)

2. (–3,6) and (1,8)

3. (–9,2) and (–1,–6)

4. (8,–5) and (0,–5)

5. (1,–3) and (–8,3)

6. Find the second endpoint of a line segment that has one endpoint at (–5,8) and a midpoint at (3,–2).

ANSWERS

Practice 1

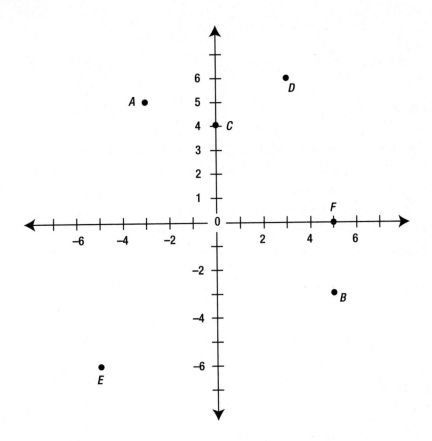

1. Quadrant II
2. Quadrant II
3. Quadrant II
4. Quadrant II
5. Quadrant II
6. Quadrant II
7. Quadrant II
8. Quadrant IV
9. y-axis
10. Quadrant I
11. Quadrant III
12. x-axis
13. (–4,2)
14. (1,–5)

15. (4,0)

16. (–6,–4)

17. (–2,6)

18. (0,–2)

19. (6,–1)

20. (3,4)

Practice 2

1. The distance formula comes from the Pythagorean theorem.
2. False: You must find the sum inside the parentheses before taking the square root of the terms.
3. False: Unless the coordinate pairs have the same x- or y-coordinate, the distance between them will be a diagonal line, which cannot be counted by hand.
4. $d = 4$
5. $d = 10$
6. $d = 5$
7. $d = 13$

Practice 3

1. (–4,8)
2. (–1,7)
3. (–5,–2)
4. (4,–5)
5. (–3.5,0)
6. midpoint $= \left(\dfrac{x_1 + x_2}{2}, \dfrac{y_1 + y_2}{2} \right)$

$(3,-2) = \left(\dfrac{-5 + x_2}{2}, \dfrac{8 + y_2}{2} \right)$

$(3,-2)$ must equal $(\frac{6}{2}, \frac{-4}{2})$

So working backwards you can reason:

$(3,-2) = (\frac{-5 + 11}{2}, \frac{8 + -12}{2})$

endpoint $= (11,-12)$

slope of a line

The things of this world cannot be made known
without a knowledge of mathematics.
—Roger Bacon

In this lesson you will learn what the slope of a line is and how slope is measured. You will learn about the slopes of horizontal, vertical, parallel, and perpendicular lines.

WHAT IS SLOPE?

Any two points on a coordinate grid can be connected to form a line that will have a slope. Similar to how different ski runs at a winter resort have different trails of varying steepness, so do the lines on a coordinate grid. The **slope** of a line is its measure of steepness. Lines are read like words—from left to right. If a line is angled upward as you look at it from left to right, then it has a *positive slope*. Conversely, if a line is angled downward when read from left to right, it has a *negative slope*. In the next figure, line *MJ* has a *steep positive slope* and line *SF* has a *moderate negative slope*.

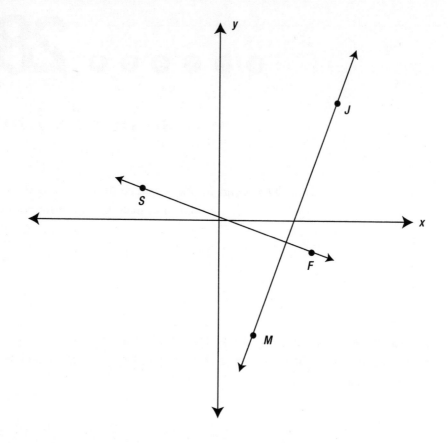

HOW IS SLOPE MEASURED?

Sometimes, people have different opinions of what is steep. One runner might think that the hill on Palms Avenue is terribly steep, but another runner might think it is only a moderate hill. When we are talking about the steepness of lines, we must use a specific and uniform method to measure slope. By exact definition, slope is the ratio of the vertical change between two points divided by the horizontal change between those same two points. This is commonly said, "change in y, over change in x." An accepted shorthand way of writing slope is with the Greek symbol delta (Δ), which means change: $\frac{\Delta y}{\Delta x}$. The letter that is always used to represent slope in coordinate geometry is m.

TIP: The slope, m, between any two points (x_1, y_1) and (x_2, y_2) is $m = \frac{y_2 - y_1}{x_2 - x_1}$.

Another way that people refer to slope is $\frac{\text{rise}}{\text{run}}$, which is said "rise over run." This is a helpful method to remember that the vertical change, or y-shift, is in the numerator of the slope, and that the horizontal change, or x-shift, is in the denominator. The most common mistake that students make is subtracting the x-coordinates in the numerator of the fraction.

··

TIP: Avoid the common mistake of putting the x-coordinates in the numerator by always thinking of $\frac{\text{rise}}{\text{run}}$ when working with slope.

··

Example: Find the slope between the points (–3,7) and (9,4).

Solution: Let (–3,7) be (x_1, y_1); and let (9,4) be (x_2, y_2).

$$m = \frac{y_2 - y_1}{x_2 - x_1}$$

$$m = \frac{4 - 7}{9 - -3}$$

$$m = \frac{-3}{9 + 3} = \frac{-3}{12} = \frac{-1}{4}$$

A slope of $m = \frac{-1}{4}$ means that for every one unit down, the line moves 4 units to the right. Since slope represents a ratio of change between the y-coordinates and the x-coordinates, it is best to keep it in fractional form. Improper fractions are fine, but mixed fractions should never be used for slope, nor should decimals. Last, when you're writing a negative slope, the negative sign is typically written in front of the fraction or with the numerator, but these three slopes all have the same meaning: $\frac{-1}{2}, \frac{1}{-2}, -\frac{1}{2}$

··

TIP: Slope should be written in simplest fractional form, never in mixed fractions, nor in decimals.

··

PRACTICE 1

1. True or false: A line can have a negative slope or positive slope, depending on which way you look at it.

2. True or false: Slope is the change in x-coordinates divided by the change in y-coordinates.

3. What are two abbreviated ways to represent slope, other than the algebraic formula?

4. Which of the following slopes is different from the rest?
   a. $\frac{-4}{1}$
   b. $\frac{4}{-1}$
   c. -4
   d. $-\frac{4}{1}$

For questions **5–8**, find the slope between the two pairs of points.

5. $(0,5)$ and $(-1,7)$

6. $(9,6)$ and $(-3,-4)$

7. $(-2,-9)$ and $(7,-6)$

8. $(-7,13)$ and $(-4,4)$

THE SLOPE OF VERTICAL LINES

Both vertical lines and horizontal lines have special slopes. When two different points have the same x-coordinate, the line connecting them is vertical. Look at the two points in the next figure and what the resulting slope is.

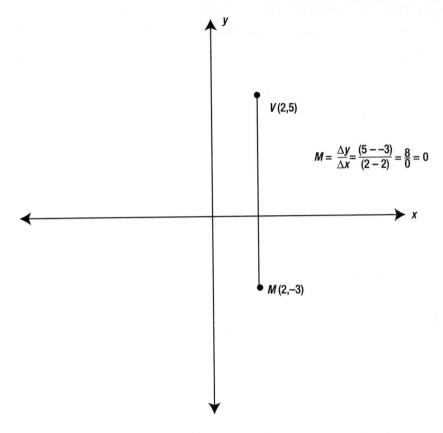

$$M = \frac{\Delta y}{\Delta x} = \frac{(5 - -3)}{(2 - 2)} = \frac{8}{0} = 0$$

As you can see, the change in the y-coordinates is 8, but the change in the x-coordinates is 0. Any fraction that has zero in the denominator is undefined, since it is impossible to divide any number by zero. (Can zero people divide 8 brownies? No! There is no one to eat the brownies, so this situation is undefined!) In order to remember that a vertical line has an undefined slope, I imagine walking down a ski slope that goes straight down. This would be impossible to do, so I know that a vertical line's slope is undefined.

....

TIP: Vertical lines have an undefined slope.

....

THE SLOPE OF HORIZONTAL LINES

Next, let us investigate the slope of a horizontal line. In this case, any two points on the line will have the same y-coordinate. Look at the two points in the next figure.

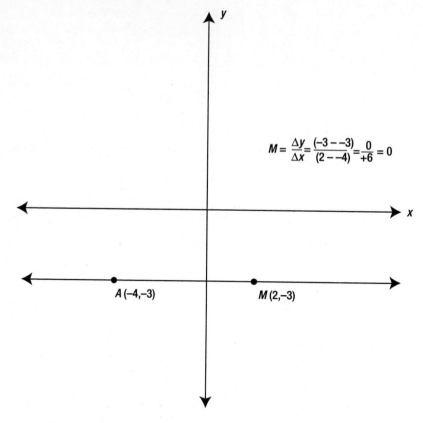

$$M = \frac{\Delta y}{\Delta x} = \frac{(-3--3)}{(2--4)} = \frac{0}{+6} = 0$$

As you can see, the change in the x-coordinates is 6, but the change in the y-coordinates is 0. Any fraction that has zero in the numerator is equal to zero; if zero brownies are shared by 6 people, every person gets zero brownies. I still think of the ski slope likeness to help me remember that horizontal lines have a slope of zero: If I were to walk on a completely flat trail, I would be able to tell a friend that it had no incline or decline.

TIP: Horizontal lines have a slope of zero.

PRACTICE 2

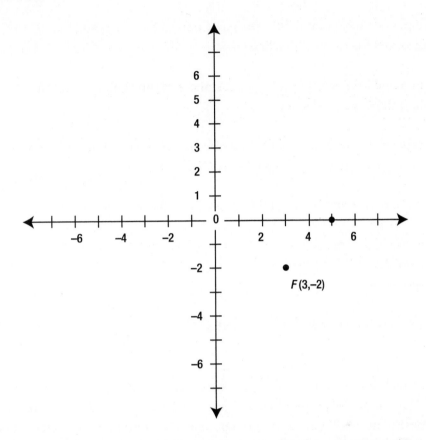

Starting with point *F* in the coordinate plane provided in this figure, plot and label points *A* through *H* so that the requirements for each question are met. (Do not use the origin as one of your points until you are instructed to do so in question **6**.)

1. Plot and label point *A* in Quadrant III, so that the line connecting *A* and *F* has a negative slope.

2. Plot and label point *B* in Quadrant I, so that the line connecting *B* and *F* has a negative slope.

3. Plot and label point *C* in Quadrant I, so that the line connecting *C* and *F* has a positive slope.

4. Plot and label point *D*, so that the line connecting *D* and *F* has an undefined slope.

5. Plot and label point *E*, so that the line connecting *E* and *F* has a slope of zero.

6. Plot and label point *G* at the origin. Connect the line going from *G* to *F* and find the slope of this line.

7. Plot and label point *H* at (–4,–6). Connect the line going from *G* to *H*, and find the slope of this line.

8. What do you notice about the intersection of the two lines from questions **6** and **7**? What kind of angle do they seem to form?

9. What do you notice about the slopes from questions **6** and **7**?

10. If you had to make up a rule for the slopes of perpendicular lines, what would it be?

SLOPES OF PERPENDICULAR LINES

Two lines are perpendicular when they form right angles. In the preceding three practice questions, you began to investigate the slopes of perpendicular lines. The **slopes of perpendicular lines** will always be **opposite reciprocals**. "Opposite" means having opposite signs and a "reciprocal" is when the numerator and denominator of a fraction switch places. For example, the opposite reciprocal of $\frac{c}{d}$ will be $\frac{-d}{c}$. When you're looking for opposite reciprocals, remember that "–3" is equivalent to $\frac{-3}{1}$, $\frac{3}{-1}$, or $-\frac{3}{1}$. Always reduce slopes into simplest terms before identifying their relationship to one another.

> **TIP:** Perpendicular lines have slopes that are opposite reciprocals.

SLOPES OF PARALLEL LINES

When two lines are parallel, it means they will never intersect. The reason they will not intersect is because they have the same rate of change. Therefore, the *slopes of parallel lines* will always be *equal*.

TIP: Parallel lines have slopes that are the same.

In the next figure, look at the examples of perpendicular and parallel lines and how their slopes are related.

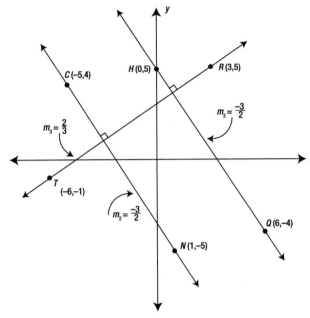

- m_1 between C and $N = \frac{4--5}{-5-1} = \frac{+9}{-6} = \frac{-3}{2}$
- m_2 between H and $Q = \frac{5--4}{0-6} = \frac{+9}{-6} = \frac{-3}{2}$

  Since $m_1 = m_2$, $\overline{CN} \parallel \overline{HQ}$

- m_3 between T and $R = \frac{-1-5}{-6-3} = \frac{-6}{-9} = \frac{2}{3}$

  Since m_3 is an opposite reciprocal to m_1 and m_2, $\overline{TR}$ is perpendicular to $\overline{CN}$ and $\overline{TR}$ is perpendicular to $\overline{HQ}$

PRACTICE 3

m_1 and m_2 represent the slopes of two different lines. Identify whether these lines are parallel, perpendicular, or neither, and explain your answer.

1. $m_1 = \frac{-4}{7}$; $m_2 = \frac{7}{-4}$

2. $m_1 = \frac{0}{-3}$; $m_2 = \frac{0}{2}$

3. $m_1 = 6$; $m_2 = \frac{1}{6}$

4. $m_1 = \frac{-12}{4}$; $m_2 = \frac{3}{9}$

5. $m_1 = -9$; $m_2 = \frac{9}{-1}$

6. $m_1 = \frac{1}{2}$; $m_2 = -2$

7. $m_1 = \frac{0}{5}$; $m_2 = \frac{8}{0}$

ANSWERS

Practice 1

1. False: Lines are read from left to right and cannot have both a positive and a negative slope.
2. False: Slope is the change in *y-coordinates* divided by the change in *x-coordinates.*
3. $\frac{\Delta y}{\Delta x}$ and $\frac{\text{rise}}{\text{run}}$ are two abbreviated ways to represent slope.
4. None of the listed slopes is different from the rest. Choices **a** through **d** all represent a slope of –4.
5. $m = \frac{y_2 - y_1}{x_2 - x_1} = \frac{5-7}{0--1} = \frac{-2}{1} = -\frac{2}{1} = \frac{-2}{1}$
6. $m = \frac{y_2 - y_1}{x_2 - x_1} = \frac{6--4}{9--3} = \frac{10}{12} = \frac{5}{6}$
7. $m = \frac{y_2 - y_1}{x_2 - x_1} = \frac{-9--6}{-2-7} = \frac{-3}{-9} = \frac{1}{3}$
8. $m = \frac{y_2 - y_1}{x_2 - x_1} - \frac{13-4}{-7--4} = \frac{9}{-3} = \frac{-3}{1}$

Practice 2

1. Any point *A* with a negative *x*-coordinate and a *y*-coordinate less than 0 and greater than –2 will be correct.
2. Any point *B* with an *x*-coordinate greater than 0 and less than 3 and a positive *y*-coordinate will be correct.
3. Any point *C* with an *x*-coordinate greater 3 and a positive *y*-coordinate will be correct.
4. Any point *D* with an *x*-coordinate of 3 will be correct.
5. Any point *E* with a *y*-coordinate of –2 will be correct.
6. Point *G* will be at (0,0) and the slope of *GF* will be $\frac{-2}{3}$.
7. The slope of *GH* will be $\frac{-6}{-4} = \frac{3}{2}$.
8. *GH* and *GF* are perpendicular and form a right angle.
9. The slopes of *GF* and *GH* are opposite signs and are reciprocals of each other.
10. If two slopes are opposite reciprocals, then the lines are perpendicular.

Practice 3

1. Neither: the slopes are reciprocals, but both have the same sign
2. Parallel: both of these slopes equal 0
3. Neither: the slopes are reciprocals, but both have the same sign
4. Perpendicular: $m_1 = \frac{-12}{4} = \frac{-3}{1}$; and $m_2 = \frac{3}{9} = \frac{1}{3}$, so they are opposite reciprocals
5. Parallel: both of these slopes equal -9
6. Perpendicular: $m_1 = \frac{1}{2}$; and $m_2 = \frac{-2}{1}$, so they are opposite reciprocals
7. Perpendicular: $m_1 = 0$, so it is a horizontal line; m_2 is undefined, so it is a vertical line

introduction to linear equations

I'm sorry to say that the subject I most disliked was mathematics.
I have thought about it. I think the reason was
that mathematics leaves no room for argument.
If you made a mistake, that was all there was to it.
—MALCOLM X

In this lesson you will learn how to identify a linear equation. You will also learn how to see if a point lies on a line and how to graph a line when you're given a linear equation.

WHY ARE LINEAR EQUATIONS IMPORTANT?

Linear equations are used to model many real-life situations. Consider placing an online order of T-shirts. There may be a constant shipping cost that applies to all orders. On top of that shipping charge, with every T-shirt ordered your bill total will increase at a constant rate. A linear equation will show exactly how the total cost of your order increases with the number of T-shirts you purchase. Linear equations are helpful for making predictions and, when graphed, help us understand the relationship between two things.

WHAT DOES A LINEAR EQUATION LOOK LIKE?

Linear equations have a standard form $Ax + By = C$, where A, B, and C are all numbers and x and y remain as variables. When you're trying to determine if an equation is linear, the signs of the terms do not matter, nor does the order that the terms occur in the equation. An equation can still be linear if the y-term comes before the x-term, or if the y-term is on the opposite side of the equation from the x-term. Sometimes, A or B will be zero and will cancel out the x or y variable. This is okay and the resulting equation will still be linear. Examples of other forms a linear equation can take are:

$$Ax + C = By$$

$$-By - Ax = C$$

$$C = Ax \text{ (in this case, } B = 0)$$

$$Ax = -By \text{ (in this case, } C = 0)$$

$$y = -5 \text{ (in this case, } A = 0 \text{ and } C = -5)$$

An important factor in linear equations is that x and y do not have exponents. Any equation with x^2 or y^2 will *not* be a linear equation. Also, x and y cannot be multiplied by each other in linear equations. Take a look at the next figure for examples of linear and nonlinear equations.

Linear Equations

$2x + 3y = 7$

$x = -3$

$y = \frac{2}{3}x + 4$

$y = 7$

Nonlinear Equations

$2x^2 + 3y = 7$

$\frac{-3}{y} = x$

$y^2 = \frac{2}{3}x + 4$

$xy = 7$

TIP: The standard form of a linear equation is $Ax + By = C$, where A, B, and C are all constant numbers and x and y are variables.

PRACTICE 1

State whether each equation represents a linear or a nonlinear equation.

1. $4x + 8 = -3y$

2. $x = 7y^2$

3. $\frac{3}{y} = -2x$

4. $x^2 + 4 = 8y$

5. $-9y - \frac{2}{3}x = 0$

6. $\frac{7}{x} = 9$

7. $y = 0.5$

8. $x \div y = 4$

GRAPHING LINEAR EQUATIONS

In order to graph a linear equation, you first need to find three points that each make the equation work. Although only two points are *needed* to define a line, finding a third point helps confirm that there were no arithmetic mistakes made while you were finding the points. (If the three points do not sit on the same line, then an error has been made.) Therefore, three points should always be used. After finding three points, plot them on a coordinate grid. As you learned in the last section, linear equations can be presented in many different ways. What is the easiest way to find points that will work as a linear equation? First, isolate the y variable in the equation. Look at the following example, where the equation $-6x + 3y = 9$ is manipulated using algebraic principles, to get y on its own.

> **Example:** Isolate y in the following equation: $-6x + 3y = 9$
>
> **Solution:**
>
> $$-6x + 3y = 9$$
> $$\underline{+6x \qquad\quad + 6x}$$
> $$3y = 9 + 6x$$
> $$\underline{\div 3 \qquad\quad \div 3}$$
> $$y = 3 + 2x$$

The line $y = 3 + 2x$ is the same as the original line, $6x + 3y = 9$, but it is just in a different format. Once y is by itself, it is easy to find three points that will work in the equation. Simply plug in three different values for x and then evaluate the expression $3 + 2x$ for y. (It is usually best to plug in x-values between -3 and 3. Then your points will be easier to graph, since they will not be too far from the origin.)

Column 1	Column 2	Column 3
let $x = 0$	let $x = 1$	let $x = -2$
$y = 3 + 2x$	$y = 3 + 2x$	$y = 3 + 2x$
$y = 3 + 2(0)$	$y = 3 + 2(1)$	$y = 3 + 2(-2)$
$y = 3$	$y = 5$	$y = -1$

You now know that the following three points will lie on the line $-6x + 3y = 9$: (0,3); (1,5); and (−2,−1). To graph the line, plot these points on a graph and connect them. In the next figure the points have been plotted for you. Now connect them to make the line for $-6x + 3y = 9$. (Remember that this can also be called line $y = 3 + 2x$.)

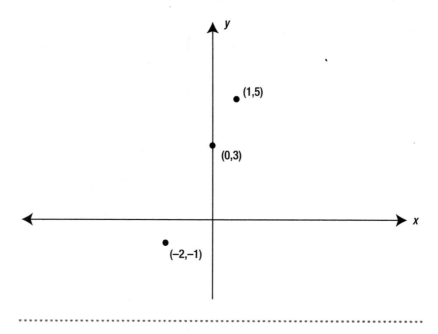

....

TIP: To graph a linear equation:

1. Get the y variable by itself.
2. Substitute three different values for x and solve for y in order to find three coordinate pairs that satisfy the equation.
3. Plot the three coordinate pairs on a graph and connect them.

....

One tricky thing to look out for is that often the coefficient of x will be in fractional form, such as in $y = \frac{3}{4}x - 2$. When this is the case, use values for x that are divisible by the denominator of the fraction being multiplied by x. It is also helpful to use zero for x.

> **Example:** What x-values would be easiest to solve for in the linear equation $2x - 3y = -6$?
>
> **Solution:**

$$2x - 3y = -6$$
$$\underline{-2x \qquad\qquad -2x}$$
$$-3y = -6 - 2x$$
$$\underline{\div -3 \qquad\quad \div -3}$$
$$y = 2 + \tfrac{2}{3}x$$

In this case, you would want to pick values for x that will be divisible by 3, such as $x = 0, 3, 6, -3, -6, 9$ and so on.

PRACTICE 2

For questions **1–4**, use algebra to get the y-variable by itself.

1. $12 = -4x + 3y$

2. $9y - \tfrac{2}{3}x = 6$

3. $7x = 5y - 10$

4. $\tfrac{1}{2}y + (-6) = x$

For questions **5–8**, find three points that would work in the given linear equation.

5. $3y - 6x = -12$

6. $2x = 3y - 3$

7. $7y + 7x = -14$

8. $\tfrac{2}{3}y = x + \tfrac{10}{3}$

DETERMINING IF POINTS ARE ON LINEAR EQUATIONS

Now that you know how to find points on a line and how to graph linear equations, this next task will be easy for you. Sometimes, you will be asked if a particular coordinate pair sits on a given line, or "satisfies the equation." In order to do this, simply plug the x and y values into the linear equation, and see if the result is a true statement. You will be happy to learn that for this method, you do not have to change the format of the line.

> **Example:** Do the points $(-4,-10)$ and $(5,4)$ satisfy the line $-3x + 2y = -8$?
>
> **Solution:**
>
> Plug in $(-4,-10)$ Plug in $(5,4)$
> $-3(-4) + 2(-10) = -8$ $-3(5) + 2(4) = -8$
> $12 + -20 = -8$ $-15 + 8 = -8$
> $-8 = -8$ $-7 \neq -8$
>
> This is a true statement, so   Since -7 does not equal -8,
> the point $(-4,-10)$ satisfies   the point $(5,4)$ does not satisfy
> the equation and must lie   the equation and does *not* lie
> on the line $-3x + 2y = -8$.   on the line $-3x + 2y = -8$.

EQUATIONS FOR HORIZONTAL AND VERTICAL LINES

Horizontal and vertical lines have unique equations that at first might not look like they are in the standard form $Ax + By = C$. With a horizontal line, the y-coordinate will be the same for every point, and only the x-coordinate will be changing. That is why horizontal lines have the form $y = C$. The x-coordinate can be an infinite number of values without affecting the y-value, so x does not need to be in the equation.

....

TIP: Horizontal lines are written in the form $y = C$, where C is any constant number.

....

On the other hand, vertical lines have a constant x-value. The value of y does not affect the value of x in vertical lines, so y does not need to be in the equation for vertical lines. Vertical lines have a standard form of $x = C$.

TIP: Vertical lines are written $x = C$, where C is any constant number.

PRACTICE 3

Identify whether the coordinate pairs in questions **1–3** satisfy the linear equation $2x - 5y = 10$.

1. $(3,-1)$

2. $(\frac{5}{2},-1)$

3. $(-5,4)$

Identify where the coordinate pairs in questions **4–6** satisfy the linear equation $\frac{1}{2}y - \frac{3}{2}x = 4$.

4. $(0,8)$

5. $(1,11)$

6. $(2,-2)$

Identify whether the linear equations in questions **7–10** represent horizontal, linear, or sloped lines.

7. $y = -7$

8. $2 = 3x$

9. The line between points $(-5,0)$ and $(0,-5)$

10. The line between points $(-2,8)$ and $(-2,-7)$

ANSWERS

Practice 1

1. linear

2. nonlinear (cannot have y^2)

3. nonlinear (cannot have yx, which is what will happen when both sides are multiplied by y)

4. nonlinear (cannot have x^2)

5. linear

6. linear (will be $7 = 9x$ when both sides are multiplied by x)

7. linear

8. linear (In order to change $x \div y = 4$ into a standard linear format, multiply both sides by y to get $x = 4y$, and then divide by 4 to end with $y = \frac{1}{4}x$.)

Practice 2

1. $y = 4 + \frac{4}{3}x$

2. $y = \frac{2}{3} + \frac{2}{27}x$

3. $y = \frac{7}{5}x + 2$

4. $y = 2x + 12$

5. Any points that work in: $y = 2x - 4$. Examples include: $(0,-4)$; $(1,-2)$; $(2,0)$; $(-1,-6)$

6. Any points that work in: $y = \frac{2}{3}x + 1$. Examples include: $(0,1)$; $(3,3)$; $(6,5)$; $(-3,-1)$

7. Any points that work in: $y = -2 - x$. Examples include: $(0,-2)$; $(1,-3)$; $(2,-4)$; $(-1,-1)$

8. Any points that work in: $y = \frac{3}{2}x + 5$. Examples include: $(0,5)$; $(2,8)$; $(4,11)$; $(-2,2)$

Practice 3

1. No. $11 \neq 10$

2. Yes. $10 = 10$

3. No. $-30 \neq 10$

For questions **4–6**, make sure you substituted the x- and y-values into the equation correctly—the order of x and y in the given equation $\frac{1}{2}y - \frac{3}{2}x = 4$ was flipped.

4. Yes. $4 - 0 = 4$

5. Yes. $5.5 - 1.5 = 4$

6. No. $-2 - 2 \neq 4$

7. $y = -7$ is horizontal

8. $2 = 3x$ is the same as $x = \frac{2}{3}$, so it is _____.

9. The line between points $(-5,0)$ and $(0,-5)$ is sloped.

10. The line between points $(-2,8)$ and $(-2,-7)$ is vertical with an equation,

$x = -2$.

transformations: reflections, translations, and rotations

There is no excellent beauty that has not some strangeness in the proportion.
—SIR FRANCIS BACON

In this lesson you will learn about various geometric transformations such as reflections, translations, and rotations.

WHAT ARE GEOMETRIC TRANFORMATIONS?

When something is "transformed" it is changed in one way or another. In geometry, points, lines, and even entire polygons can be transformed on a coordinate grid. Sometimes, a transformation will result in movement in one or two directions, but the size and appearance of the original figure will remain the same. Some transformations will create a mirror image of an object, and other transformations will change a figure's size.

HOW ARE TRANSFORMATIONS NAMED?

Regardless of which type of transformation is being used, all transformations use the same method for naming the original image and the transformed image. The original image is called the **preimage** and the new figure is called the **image**. Points in transformations have a specific way of being named. When a point undergoes a transformation, it keeps the same name, but it is followed by a prime mark ('). For example, after point *Q* is transformed, it becomes *Q'*, which is said, "*Q* prime." Look at the transformation of polygon *ABCD* in the next figure. Notice that the image it creates is named *A'B'C'D'*.

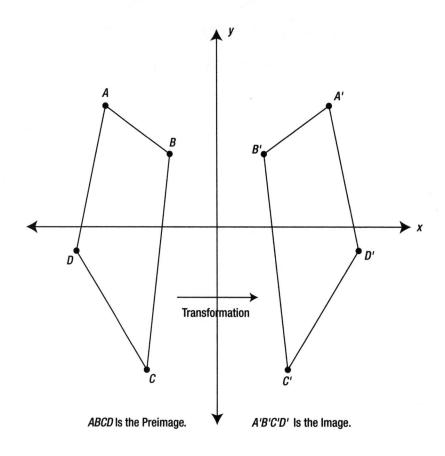

ABCD Is the Preimage.   *A'B'C'D'* Is the Image.

· ·

TIP: A transformation is a shift of a *preimage* into a new figure called the *image*. Transformed points are named with the same variable as the original point, but are followed with a prime mark (').

· ·

WHAT IS A REFLECTION?

A **reflection** is a transformation that flips a point or object over a given line. While the shape of the object will be the same, it will be a mirror image of itself. It is common that points, lines, or polygons will be reflected over the *x*-axis or *y*-axis, but they can be reflected over other lines as well. Polygon *ABCD* in the preceding figure underwent a reflection over the y-axis to become $A'B'C'D'$. In order to reflect a point over the y-axis, its y-coordinate stays the same, and its x-coordinate changes signs. Point $A(x,y)$ will become $A'(-x,y)$ after being reflected over the y-axis.

TIP: Point $A(x,y)$ reflected over the y-axis becomes $A'(-x,y)$.

Similarly, in order to reflect a point over the x-axis, its x-coordinate will stay the same, and its y-coordinate will change signs. Point $A(x,y)$ will become $A'(x,-y)$ after being reflected over the x-axis.

TIP: Point $A(x,y)$ reflected over the x-axis will become $A'(x,-y)$.

This figure demonstrates the reflection of point P over the x-axis.

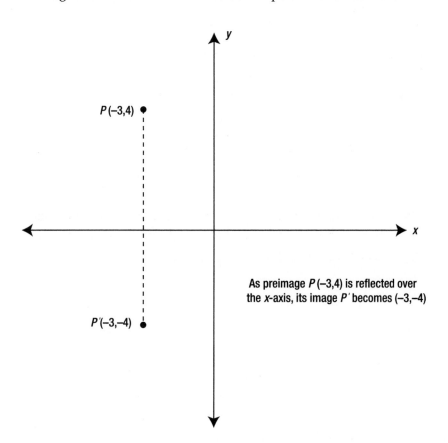

As preimage $P(-3,4)$ is reflected over the x-axis, its image P' becomes $(-3,-4)$

WHAT IS A TRANSLATION?

A **translation** is a transformation that slides a point or object horizontally, vertically, or in both directions, which creates a diagonal shift. When a figure is translated, it will look exactly the same, but will just have a different location.

TIP: When a point is *translated*, it is shifted horizontally, vertically, or diagonally. Translated figures keep the same appearance and size.

The notation $T_{h,k}$ is used to give translation instructions. The subscript h shows the movement that will be applied to the x-coordinate and the subscript k shows the movement that will be applied to the y-coordinate. These numbers are each added to the x-coordinate and y-coordinate in the preimage.

TIP: Point $A(x,y)$ translated using $T_{h,k}$, becomes $A'((x + h), (y + k))$.

Example: Translate $V(4,-5)$ using translation $T_{-2,4}$.

Solution: Beginning with $V(4,-5)$, add -2 to the x-coordinate and add 4 to the y-coordinate: V-prime would be $V'((4 + -2), (-5 + 4)) = V'(2,-1)$. This translation is illustrated in the next figure.

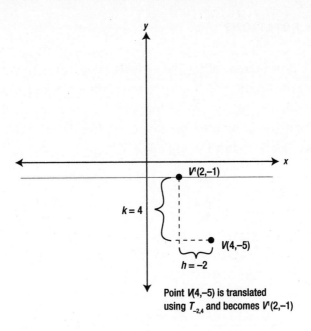

Point V(4,–5) is translated
using $T_{-2,4}$ and becomes V'(2,–1)

PRACTICE 1

1. After transforming point W, the new point will be named _____.

2. True or false: A transformation will always change the size of an object.

3. True or false: A translation will result in an image that is a mirror of the preimage.

4. How do you translate a point $R(x,y)$ using $T_{h,k}$?

5. After reflecting $N(-3,7)$ over the x-axis, what will the coordinates of N' be?

6. After reflecting $M(6,-1)$ over the y-axis, what will the coordinates of M' be?

7. After translating $L(2,-5)$ using $T_{-3,-4}$, what will the coordinates of L' be?

8. Reflect point $J(5,-2)$ over the x-axis to get point J'. Then reflect J' over the y-axis to get J''. What will the coordinate be of J''?

WHAT IS A ROTATION?

A rotation is a type of translation that swivels a point or polygon around a fixed point, such as the origin. Rotations are done either clockwise or counterclockwise and are done in degrees. Most commonly, figures are rotated 90° or 180°. In the next figure, you can observe how triangle *KAY* in Quadrant I is rotated counterclockwise 90° to form triangle *K'A'Y'*.

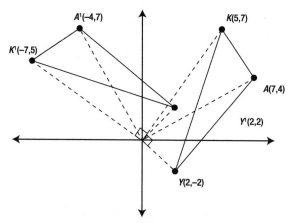

Can you notice what is happening each time a point
is rotated 90° counterclockwise?

90° counterclockwise
K(5,7) ⟶ K'(−7,5)
A(7,4) ⟶ A'(−4,7)
Y(2,−2) ⟶ Y'(2,2)

By looking at what happens to points *K*, *A*, and *Y* in this figure as they are rotated counterclockwise 90°, can you guess what the rule is for this type of translation? Did you notice that the preimage *x*-coordinate becomes the image *y*-coordinate? Also, the preimage *y*-coordinate switches signs and then becomes the *x*-coordinate in the image point. This relationship is summed up as follows:

TIP: When preimage point *A(x,y)* is *rotated 90° counterclockwise* around the origin, the image point will be *A'(−y,x)*.

Look at the next figure to see what happens to points that are rotated 90°clockwise.

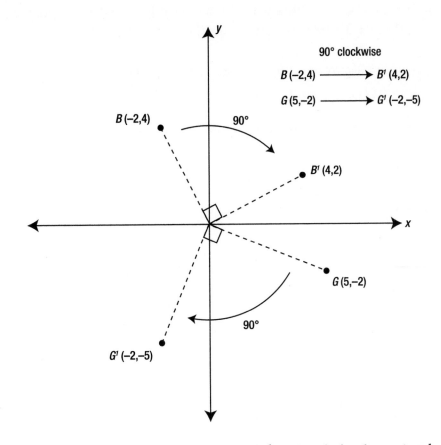

In this case, the preimage x-coordinate switches signs before becoming the image y-coordinate. The preimage y-coordinate becomes the x-coordinate in the image point.

....

TIP: When preimage point $A(x,y)$ is *rotated 90° clockwise* around the origin, the image point will be $A'(y,-x)$.

....

TRANSFORMING POLYGONS

You have seen a few examples of entire polygons undergoing transformations. To reflect, translate, or rotate an entire polygon, apply the appropriate transformation rule to each individual coordinate pair in the polygon, as was done to triangle *KAY* in the preceding figure about *KAY*.

PRACTICE 2

1. True or false: To rotate a point around the origin, switch the order of x and y and change both of their signs.

2. True or false: When $U(8,-3)$ is rotated 90° clockwise, the result will be $U'(-3,-8)$.

For questions **3–8**, identify the image point bases on the transformation given:

3. Preimage: $T(-3, 5)$; Rotate 90° counterclockwise

4. Preimage: $S(4,-2)$; Rotate 90° clockwise

5. Preimage: $I(2,7)$; Reflect over the x-axis

6. Preimage: $F(-1,-3)$; Reflect over the y-axis

7.. Preimage: $U(-1,-3)$; Translate using $T_{-6,5}$

8. Preimage: $Z(9,-3)$; Reflect over the y-axis, then rotate 90° counterclockwise to find ″

ANSWERS

Practice 1

1. After transforming point W, the new point will be named W'.
2. False: Only some types of transformations change the size of an object.
3. False: Reflections result in mirror images, but translations do not.
4. $R'((x + h),(y + k))$
5. $N'(-3,-7)$
6. $M'(-6,-1)$
7. $L'((2 + -3), (-5 + -4)) = L'(-1,-9)$
8. $J'(5,2)$ and $J''(-5,2)$

Practice 2

1. False: The x- and y-coordinates will switch places, but only one of them will change signs. Which one depends on whether the rotation is clockwise or counterclockwise.
2. True.
3. $T'(-5,3)$
4. $S'(-2,4)$
5. $I'(2,-7)$
6. $F'(1,-3)$
7. $U'(-7,2)$
8. $Z''(3,-9)$

GLOSSARY

adjacent angles angles that share a common vertex, have one common side, and have no interior points in common

alternate exterior angles When a transversal crosses a pair of lines, the alternate exterior angles are the angles outside the pair of lines, on opposite sides of the transversal. If the two lines crossed by the transversal are parallel, then the alternate exterior angles will be congruent.

alternate interior angles When a transversal crosses a pair of lines, the alternate interior angles are the angles inside the pair of lines, on opposite sides of the transversal. If the pair of lines crossed by the transversal is parallel, then the alternate interior angles will be congruent.

altitude of a triangle the line that passes through the vertex of a triangle, forming a right angle with the opposite side, which is defined as the base of that triangle. All triangles have three altitudes.

angle bisector a line, ray, or segment that goes through the vertex of an angle, dividing it into two angles of equal measure

arc   a curved section of the outside of the circle that is defined by two points on the circle

base of a prism   the two congruent ends of a prism

bisector   a line, ray, or segment that goes through a line segment's midpoint, dividing the segment into two equal parts

central angle   an angle formed by two radii in a circle. Its vertex is the center of the circle and its endpoints sit on the circle.

chord   any line segment in a circle that extends from one point on the circle to another point on the circle

circumference   the distance around a circle, defined by $2\pi r$ or πd

complementary angles   two angles whose measures sum to 90°

coordinate plane   a plane that contains a horizontal number line (the x-axis) and a vertical number line (the y-axis). Every point in the coordinate plane can be named by a pair of numbers written in the form (x,y)

corresponding angles   When a transversal crosses a pair of lines, the corresponding angles are the angles at the same relative position at each intersection. If the two lines crossed by the transversal are parallel, the corresponding angles will be congruent.

decagon   a polygon with 10 sides

degree   the unit by which angles are measured. A circle contains 360 degrees.

diameter of a circle   a chord in a circle that passes through the center of the circle. Any diameter will be twice the length of a circle's radius.

equiangular polygon   a polygon with all angles equal in measure

equiangular triangle   a triangle with all angles of equal measure. Its sides will also be equal in length.

equilateral polygon   a polygon with all sides equal in length

equilateral triangle   a triangle with all sides of equal length. Its angles will also be equal in measure.

heptagon   a polygon with seven sides

hexagon   a polygon with six sides

inscribed angle   an angle in a circle whose sides are two chords within the circle, and whose vertex sits on the circle. The measure of an inscribed angle is half the arc it forms.

intercepted arc   the minor arc formed by a central angle in a circle. Its measure is equal to the measure of the central angle that forms it

isosceles trapezoid   a trapezoid with one pair of congruent sides

legs of an isosceles triangle   the two congruent sides in an isosceles triangle

legs of a trapezoid   the two non-parallel sides in a trapezoid

linear equation   an equation whose graph is a line. Linear equations have a standard form $Ax + By = C$, where A, B, and C are all numbers and x and y are variables.

major arc   the longer arc connecting any two points on a circle

median of a triangle   the line extending from the vertex of a triangle to the midpoint of the opposite side. All triangles have three medians.

median of a trapezoid   the line segment that joins the midpoints of the legs of a trapezoid

midpoint   a point that divides a line segment into two segments of equal length

mid-segment of a triangle   a segment that connects the midpoints of any two sides of a triangle

minor arc   the shorter arc connecting any two points on a circle

nonagon   a polygon with nine sides

octagon   a polygon with eight sides

origin   the point in a coordinate plane where the x-axis and y-axis intersect. The coordinates of the origin are (0,0).

pentagon   a polygon with five sides

plane   a collection of points that make a flat surface, which extends continuously in all directions, like an endless wall

prism   a three-dimensional figure that has two congruent polygon ends and parallelograms as the remaining sides

pyramid   a three-dimensional shape with a regular polygon base, and congruent triangular sides that meet at a common point

Pythagorean theorem   For any right triangle, the square of the length of the hypotenuse is equal to the sum of the squares of the lengths of the legs. Written as: $a^2 + b^2 = c^2$

reflection   a transformation that flips a point or object over a given axis or line

rotation   a type of translation that swivels a point or polygon around a fixed point, such as the origin

same-side interior angles   When a transversal crosses a pair of lines, the same-side interior angles are the angles inside the pair of lines, on the same side of the transversal. If the two lines crossed by the transversal are parallel, then the same-side interior angles will be supplementary.

sector of a circle   a part of a circle bound by two radii and their intercepted arc

similar   Polygons are similar when all their corresponding sides are proportional to one another.

supplementary angles   two angles whose measures sum to 180°

surface area   the combined area of all of a prism's outward-facing surfaces

translation   a transformation that slides a point or object horizontally, vertically, or in both directions, which creates a diagonal shift. The notation $T_{h,k}$ is used to give translation instructions.

transformation   a general movement or manipulation of a point or shape on a coordinate graph. Three forms of transformations are translations, reflections, and rotations. The original point or shape is called the *preimage* and the transformed object is called the *image*.

vertex   the intersection of two sides of a polygon; also the intersection of two line segments or rays that form an angle

vertical angles   a pair of non-adjacent angles formed when two rays or lines intersect

volume   the amount of space enclosed by a three-dimensional object. Volume is expressed in cubic units.

x-axis   the horizontal number line that is the scale in a coordinate plane. Values to the right of the origin are positive and those to the left are negative.

y-axis   the vertical number line that is the scale in a coordinate plane. Values above the origin are positive and those below it are negative.

additional online practice

WHETHER YOU NEED help building basic skills or preparing for an exam, visit the LearningExpress Practice Center! On this site, you can access additional online geometry practice materials. Using the code below, you'll be able to log in and take an additional practice exam. This online practice exam will also provide you with:

- Immediate scoring
- Detailed answer explanations
- Personalized recommendations for further practice and study

Log on to the LearningExpress Practice Center by using the URL: **www.learnatest.com/practice**

This is your Access Code: **7663**

Follow the steps online to redeem your access code. After you've used your access code to register with the site, you will be prompted to create a username and password. For easy reference, record them here:

Username: _____ **Password:** _____

With your username and password, you can log in and answer these practice questions as many times as you like. If you have any questions or problems, please contact LearningExpress customer service at 1-800-295-9556 ext. 2, or e-mail us at customerservice@learningexpressllc.com

P O S T T E S T

THE 30 LESSONS you have just read have given you the skills necessary to ace the posttest. The posttest has 30 questions. The questions are similar to those in the pretest, so you can see your progress and improvement since taking the pretest.

The posttest can help you identify which areas you have mastered and which areas you need to review a bit more. Return to the lessons that cover the topics that are tough for you until those topics become your strengths.

POSTTEST

1. What is the least number of points needed to define a plane?
   a. 0
   b. 1
   c. 2
   d. 3

2.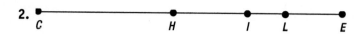

   $\overline{LE} = 6$ cm

   $\overline{IL} = \frac{2}{3}\overline{LE}$

   $\overline{HI}$ is twice as long as $\overline{IL}$

   If $\overline{CI}$ is equal to 22 cm, how long is $\overline{HI}$?

   a. 10 cm
   b. 12 cm
   c. 14 cm
   d. 16 cm

3. $\angle P$, $\angle R$, and $\angle Q$ are supplementary angles. If $m\angle Q = 120°$ and $\angle P$ is half the measure of $\angle R$, what is the $m\angle P$?
   a. 60°
   b. 30°
   c. 40°
   d. 20°

4. In the next figure, $m\angle AOT = 52°$ and $m\angle POG = 137°$. What is the $m\angle POA$?

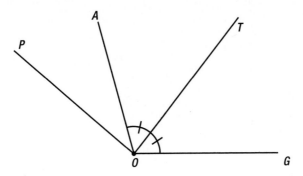

a. 33°

b. 104°

c. 52°

d. 34°

5. In the next figure, $q \parallel r$. Solve for x.

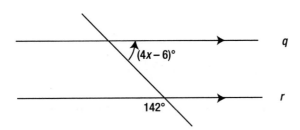

a. 38

b. 48

c. 11

d. 14

6. In $\triangle VID$, $\overline{IJ}$ could be all of the following EXCEPT:

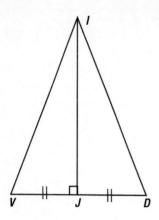

   a. Altitude
   b. Hypotenuse
   c. Perpendicular bisector
   d. Height

7. In $\triangle ABC$, find the $m\angle C$ if $m\angle A = (2x + 7)°$, $m\angle B = 28°$, and $m\angle C = (3x)°$.
   a. 29°
   b. 87°
   c. 65°
   d. 180°

8. In $\triangle PIG$, $\overline{PI} = 6$, $\overline{PG} = 8.5$, and $\overline{IG} = 6$. What kind of triangle is $\triangle PIG$?
   a. Scalene
   b. Right
   c. Isosceles
   d. Equilateral

9. Given the following triangle, which expression represents the measure of ∠Q?

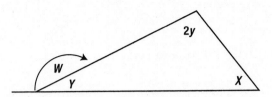

   a. $3y + x$
   b. $180° - x$
   c. $3y$
   d. $2y + x$

10. The area of $\triangle THE = 620$ cm². If the base of the triangle is 40 cm, what is the measure of the altitude of $\triangle THE$?
    a. 15.5 cm
    b. 80 cm
    c. 31 cm
    d. 62 cm

11. Find the missing side length in the given right triangle.

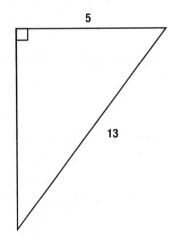

    a. 8
    b. 9
    c. 11
    d. 12

12. △*CAB* is a 45°-45°-90° triangle. If one of its legs measures 9 inches, what are the measures of the other two sides?

   a. 9 inches and $9\sqrt{2}$ inches

   b. 9 inches and $9\sqrt{3}$ inches

   c. 9 inches and 9 inches

   d. Cannot be determined with the information given

13. The two triangles that follow are congruent triangles. What is the appropriate way to name the second triangle so that it shows congruence with △*KAS*?

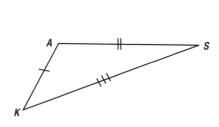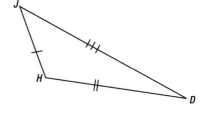

   a. △*JDH*

   b. △*DHJ*

   c. △*JHD*

   d. The order of vertices is not important when you're naming congruent triangles.

14. What is the sum of the measures of the interior angles of a pentagon?

   a. 360°

   b. 540°

   c. 720°

   d. 900°

15. Given parallelogram *ABLE* that follows, find the ∠*EAL*.

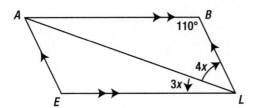

   a. 30°

   b. 40°

   c. 70°

   d. 80°

16. Which of the following statements is false?

   a. A square is a rectangle.

   b. Adjacent angles of parallelograms are supplementary.

   c. All rhombuses are parallelograms.

   d. All trapezoids have one pair of congruent sides.

17. Pete receives a box of fruits and vegetables from a local farm that contains the following items: 4 apples, 5 peppers, 2 mangoes, 3 cucumbers, 6 onions, and 1 cantaloupe. What is the ratio of fruits to vegetables that Pete received?

   a. $\frac{1}{3}$

   b. $\frac{1}{2}$

   c. $\frac{2}{3}$

   d. 2

18. The two triangles below can best be described as being

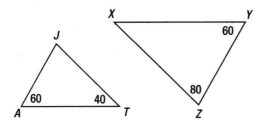

   a. Similar

   b. Congruent

   c. Regular

   d. Corresponding

19. Both a square and a regular hexagon have the same perimeter. What is the ratio of a side length of the square to a side length of the hexagon?

a. $\frac{1}{6}$

b. $\frac{1}{8}$

c. $\frac{1}{2}$

d. $\frac{2}{3}$

20. What is the area of the trapezoid that follows?

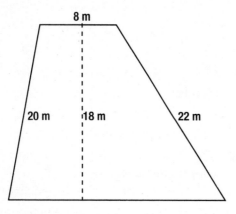

a. 357 m²

b. 378 m²

c. 468 m²

d. 306 m²

21. In the following figure, circle D has D as its center. If $\overarc{ITS} = 246°$, find the $m\angle IDS$.

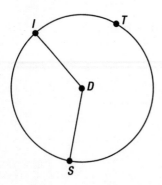

a. 57°

b. 114°

c. 123°

d. 136°

22. A circle is circumscribed within a square so that its edge touches all sides of the square, such as in the next figure. If the area of the square is 36 m², what is the circumference of the circle, in terms of π?

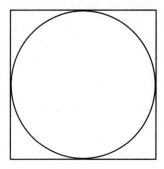

    a. 6π m
    b. 9π m
    c. 12π m
    d. 18π m

23. Using the same figure from question 22, if the perimeter of the square is 32 cm, what is the area of the circle, in terms of π?
    a. 8π cm²
    b. 64π cm²
    c. 16π cm²
    d. 32π cm²

24. The shaded area in the figure of a fountain represents a grassy area surrounding a 5-yard by 10-yard fountain. Kaoru is purchasing grass seed to plant. How many square feet of grass will she be growing in the shaded area around the pool?

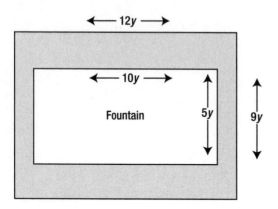

   a. 108 yd^2
   b. 50 yd^2
   c. 62 yd^2
   d. 58 yd^2

25. What is the surface area of a cylinder that has a diameter of 8 inches and a height of 14 inches? Round your answer to the nearest tenth.
   a. 418.6 in.^{2}
   b. 427.0 in.^{2}
   c. 452.2 in.^{2}
   d. 542.2 in.^{2}

26. What is the volume of the triangular prism in the next figure?

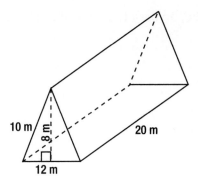

a. 2,400 m²
b. 1,640 m²
c. 1,200 m²
d. 960 m²

27. What is the distance d, between $J(-6,1)$ and $P(4,6)$?
  a. 15
  b. $5\sqrt{5}$
  c. $\sqrt{27}$
  d. $\sqrt{51}$

28. If a line has a slope of $\frac{-1}{5}$, a line *perpendicular* to it will have a slope of
  a. 5
  b. –5
  c. $\frac{1}{5}$
  d. Cannot be determined with the given information

29. Which coordinate pair is NOT on the line $\frac{2}{3}x - \frac{1}{2}y = -4$?
  a. (–3,4)
  b. (–9,–4)
  c. (3,12)
  d. (0,–8)

30. What transformation could be used to create image point $N'(-5,-9)$ from preimage point $N(9,-5)$?
  a. A reflection over the x-axis.
  b. A translation using $T_{-4,4}$.
  c. A clockwise rotation of 90°.
  d. A counterclockwise rotation of 90°.

ANSWERS

For any questions that you may have missed, go back to review the associated Lesson.

1. **d.** Lesson 1
2. **c.** Lesson 2
3. **d.** Lesson 3
4. **a.** Lesson 4
5. **c.** Lesson 5
6. **b.** Lesson 6
7. **b.** Lesson 7
8. **c.** Lesson 8
9. **d.** Lesson 9
10. **c.** Lesson 10
11. **d.** Lesson 11
12. **a.** Lesson 12
13. **c.** Lesson 13
14. **b.** Lesson 14
15. **b.** Lesson 15
16. **d.** Lesson 16
17. **b.** Lesson 17
18. **a.** Lesson 18
19. **d.** Lesson 19
20. **d.** Lesson 20
21. **b.** Lesson 21
22. **a.** Lesson 22
23. **c.** Lesson 23
24. **d.** Lesson 24
25. **c.** Lesson 25
26. **d.** Lesson 26
27. **b.** Lesson 27
28. **a.** Lesson 28
29. **d.** Lesson 29
30. **d.** Lesson 30